MATHEMATICAL MODELS IN COASTAL ENGINEERING

MATHEMATICAL MODELS IN COASTAL ENGINEERING

Christopher G. Koutitas
Professor of Civil Engineering
Aristotle University
Thessaloniki, Greece

PENTECH PRESS
London

First published 1988
by Pentech Press Limited
Graham Lodge, Graham Road
London NW4 3DG

British Library Cataloguing in Publication Data

Koutitas, Christopher G.
Mathematical models in coastal engineering.
1. Physical oceanography. Mathematical models
I. Title
551.46'01

ISBN 0-7273-1313-4

Typeset by Mid-Country Press, London SW15
Printed and bound in Great Britain

Preface

Coastal engineering has developed over the past decades to become a modern science, flourishing on a rich base of knowledge provided by Physical Oceanography, Maritime Hydraulics and Geotechnics. It has become a link between these domains and has provided a means of communication between oceanographers and civil engineers. The continuation of this interdisciplinary effort is, in my belief, an indispensable factor in the further development of coastal engineering. The purpose of this book is, therefore, to present some basic aspects of the formulation and numerical solution of mathematical models in coastal engineering as a small contribution to the major research and educational efforts being undertaken worldwide in universities and research institutes.

The use of mathematical models or more specifically of numerical models was introduced into physical oceanography and almost simultaneously to engineering hydraulics in the 1970s. This present work is particularly concerned to demonstrate the applicability of microcomputers in resolving the mathematical models which relate to coastal engineering. The aim is twofold:

(1) To encourage engineers in the smaller consulting firms, not having easy access to a main frame computer, to use microcomputers to deepen their insight into the behaviour of the small-scale coastal structures they may design and to examine the many possible alternative solutions by applying the same sophisticated methodology as is used in the design of major coastal structures.

(2) By removing the mystique, to encourage physical oceanographers to use computers in the investigation of physical coastal oceanographic processes.

The book does not attempt to go too deeply into the physical analysis of the problems which arise in coastal engineering but concentrates on the more fundamental aspects of the theory required for the formulation of mathematical models and their numerical solution. A prerequisite knowledge for an easy understanding of the book is that of the basics of coastal engineering, physical oceanographic and hydraulic principles, common to all engineering and meteorology-oceanography seniors or graduate students. In addition, a knowledge of the basic methods of numerical analysis and specifically of finite difference techniques and BASIC or FORTRAN programming is required.

The first chapter deals with the mathematical theory of waves and the statistical properties of wind generated waves; the second with coastal circulation due to the various generating mechanism, such as tide, wind density variations and waves; the third with pollutant advective diffusion and the fourth with sediment transport in the surf zone and the wider coastal domain. Each section contains a fully worked illustrative example of a coastal engineering problem together with the corresponding computer program listing, in BASIC, with explanations.

The programs presented are not intended as simply computer packages, ready to be used in 'black box' form, but are meant to be triggering mechanisms for those interested readers to develop and generalise them further.

The book was first published in the Greek language with the administrative and financial support of the Greek Ministry of Research and Technology. The translation was encouraged by fruitful discussions with my eminent colleagues, Professor B. O'Connor of the University of Liverpool and Professor D. M. McDowell. To all those who supported this project, from an initial idea to a book and to my family who endured my efforts with great patience, I hereby express my profound gratitude.

Christopher Koutitas
Thessaloniki

Contents

List of programs

1

Mathematical theory of waves

1.1 PHENOMENA—DEFINITIONS

Sea waves have various generating mechanisms but predominantly gravity waves are of the greatest interest in engineering practice. Such waves, which have the form of periodic or non-periodic disturbances of the sea surface, are the main factor to be considered in the design, construction and operation of most marine structures.

The aim of this book is to present the basic principles involved in the design and computation related to coastal structures—not too much space is devoted to the presentation of the mathematical theory of waves but some aspects are summarised in order to ensure that the book is self-contained. The reader is referred to a number of classical books on wave theory for a more complete treatment of this subject.

Waves can be considered as a sequence of successive pulses distinguished by the passage of the free surface from a relative maximum to a relative minimum elevation, characterised as a wave crest and a wave trough, respectively. Each periodic wave can be characterised by the basic elements of the wave height H (distance between crest and trough), the wave period T (time difference between two successive crests or troughs), the wave length L (distance between two successive crests or troughs) and c, the velocity of wave propagation or wave celerity.

Various physical phenomena in the marine environment, characterised by different names, are basically forms of sea waves governed by the same physical principles and describable by the same mathematical models. These are named and presented in Table 1.1 in order of their characteristic time scale.

1.2 CLASSIFICATION OF WAVES

Waves are classified and grouped into various categories, mainly for reasons of easy formulation and solution of the mathematical models and not because there are physical differences between them. The

Table 1.1

Phenomena	*Generating force*	*Time scale (period)*
Wind generated waves	Shear + pressure of wind on sea surface	0–15 s
Swell	Long-distance wind waves	0–30 s
Surf beats	Grouping of breaking waves	1–5 min
Seiches	Variations of wind speed and atmospheric pressure	1–60 min
Basin resonance	Tsunami, surf beats	1–60 min
Tsunami	Undersea earthquakes	5–60 min
Tide	Moon–sun influence on earth gravity	12–24 h
Storm surge	Wind shear + atmospheric pressure on sea	1–30 days

classification is made on the basis of the values of the physical parameters which characterise the waves; the height H, the length L, the period T and the water depth h.

Classification in terms of wave height leads to a distinction between waves of infinitesimal height and waves of finite height. The first case requires small values of $\delta = H/h$ and H/L. Linear wave theory, or Airy theory, or Stokes 1st order theory, or infinitesimal height theory is applicable. In the later case, Stokes 2nd or higher order theory, or cnoidal wave theory is applicable.

Waves can conveniently be classified into three distinct types according to the relative depth h/L:

$$\varepsilon = h/L < 0.05 \quad \text{Shallow-water (long) waves}$$

$$0.05 < h/L < 0.5 \quad \text{Intermediate waves}$$

$$h/L > 0.5 \quad \text{Deep-water (short) waves}$$

Classification in terms of the parameters h, L, H leads to the formation of the Ursell number, $U = HL^2/h^3 = \delta/\varepsilon^2$. Small values of U permits the application of Stokes wave theory of various orders. Larger values of U, as in the case of long waves of finite height, require the use of cnoidal wave theory. The classification in terms of H, h, L ($L \propto T^2$) is illustrated in Fig. 1.1.

1.3 CONCISE PRESENTATION OF SMALL-AMPLITUDE WAVE THEORY

The basic assumptions underlying small-amplitude wave theory are:

(1) Small values of the H/L and H/h ratios.

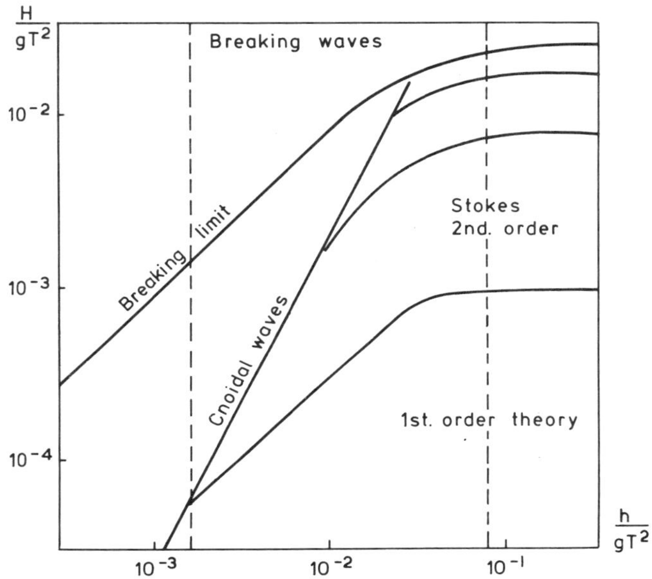

Fig. 1.1 Classification of waves and appropriate mathematical theories

(2) Constant water depth and wave period.
(3) Two-dimensionality of the phenomenon in the x–z plane, i.e. long-crested waves.
(4) Periodic waves with constant form. For progressive waves the surface elevation is approximated by the relation

$$\eta = \frac{H}{2}\cos\left(\frac{2\pi x}{L} - \frac{2\pi t}{T}\right) = \frac{H}{2}\cos(kx - \omega t) \tag{1.1}$$

(5) Incompressible fluid.
(6) Ideal fluid and no viscosity effects. This assumption leads to the conservation of vorticity for the initial (zero) value, and consequently to irrotational flow. This assumption is realistic for values of $\delta/\varepsilon = HL/h^2 < 20 - 30$. Beyond this, the boundary layer effect is important and the condition of irrotational flow is not realistic.

The mathematical model is formulated under the abovementioned assumptions in terms of the flow potential function, $\varphi(x, z, t)$

$$\mathbf{U} = (u, w) = \text{grad}\,\varphi \tag{1.2}$$

The field equation and the corresponding boundary conditions are

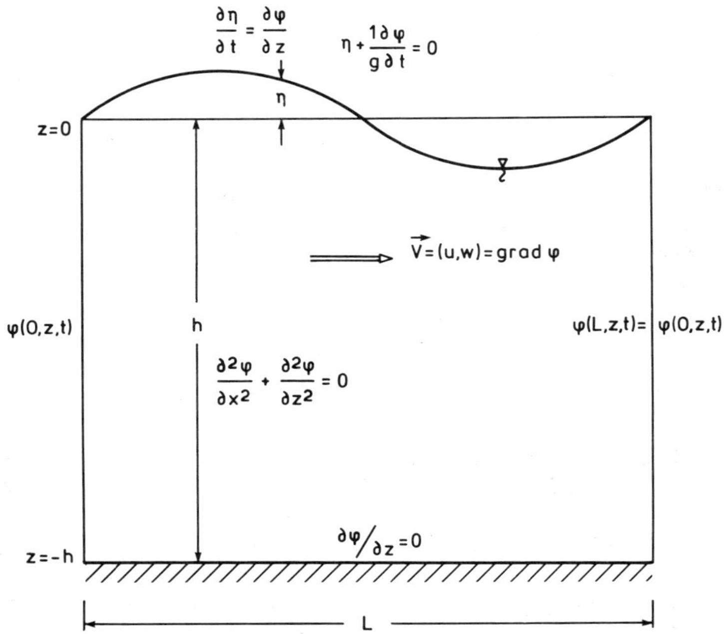

Fig. 1.2 The field equation and boundary condition of small-amplitude wave model

schematically shown in Fig. 1.2. The solution of a model consisting of the Laplace equation and the redundant free surface and bed boundary conditions (overdefined problem) leads to the following important functional relations:

(1) Free surface elevation (preassumed)

$$\eta = \frac{H}{2}\cos(kx - \omega t) \qquad \text{where } k = \frac{2\pi}{L},\ \omega = \frac{2\pi}{T} \tag{1.3}$$

(2) Velocity potential

$$\varphi = \frac{H}{2}\cdot c\cdot\frac{\cosh\{k(z+h)\}}{\cosh kh}\cdot\sin(kx - \omega t) \tag{1.4}$$

(3) Particle velocity components along Ox, Oz axes

$$u = \frac{\pi H}{T}\cdot\frac{\cosh\{k(z+h)\}}{\sinh kh}\cdot\cos(kx - \omega t) \tag{1.5}$$

$$w = \frac{\pi H}{T}\cdot\frac{\sinh\{k(z+h)\}}{\sinh kh}\cdot\sin(kx - \omega t) \tag{1.6}$$

(4) Pressure distribution over the depth

$$p = p_{\text{hydrostat}} + p_{\text{dynam}} \\ = -\rho g z + \rho g \frac{H}{2} \cdot \frac{\cosh\{k(z+h)\}}{\cosh kh} \cos(kx - \omega t) \tag{1.7}$$

(5) Frequency dispersion relation (compatibility condition). It describes the dependence of the celerity on the wave frequency (or period).

$$\omega^2 = gk \tanh kh \rightarrow L = \frac{gT^2}{2\pi} \tanh kh$$

$$\rightarrow c = \frac{gT}{2\pi} \tanh kh \tag{1.8}$$

(6) Energy content (kinetic + potential) in water volume of unit width, of length equal to L and height equal to the water depth

$$E = E_p + E_k = \frac{\rho g H^2 L}{16} + \frac{\rho g H^2 L}{16} = \frac{\rho g H^2 L}{8} = \bar{E} L \tag{1.9}$$

(where $\bar{E}$ is the energy density). The rate of flux of energy through a vertical section of unit width (wave power) is given by

$$P = \frac{E}{T} \cdot \frac{1}{2}\left(1 + \frac{2kh}{\sinh 2kh}\right) = \frac{En}{T} \tag{1.10}$$

where n is an important wave parameter.

The above relations have a general validity and take simpler forms in the limiting cases of shallow and deep waters. In the first case, the simplification is made by the approximation, $\cosh(kh) \simeq 1$, $\sinh(kh) \simeq kh$ and $\tanh(kh) \simeq kh$; in the second case by the approximation, $\tanh kh \simeq 1$. The shallow water waves or long waves are not characterised by frequency dispersion as the velocity is independent of the wave period.

$$c_{\text{shallow}} = \sqrt{(gh)} \tag{1.11}$$

1.4 WAVES OF FINITE AMPLITUDE

If the wave amplitude is more than an infinitesimal percentage of the wave length L ($O[H/L] > 0.01$) or of the water depth h ($O[H/h] > 0.1$) then the small amplitude wave theory does not approximate the phenomenon in a satisfactory manner.

In that case, when the wave length is less than the water depth (short waves) Stokes higher order theories are more accurate and when the wave length is greater than the water depth (long waves) the cnoidal waves theory is more accurate.

1.4.1 Short waves

Stokes theories of various orders make use of terms of higher than first order in the series development of φ.

$$\varphi = \varphi^{(1)} + \varphi^{(2)} + \varphi^{(3)} + \cdots \tag{1.12}$$

where $O[\varphi^{(2)}] = O[\varphi^{(1)} \cdot H/L]$, $O[\varphi^{(3)}] = O[\varphi^{(1)} H^2/L^2] \cdots$.

The basic procedure in development of Stokes theories is substitution of that series in the mathematical model, the retention of the non-linear term $\left(\dfrac{\partial\varphi}{\partial x} \cdot \dfrac{\partial\eta}{\partial x}\right)$ in the surface kinematic condition and the formulation of equations in terms of the various powers of the small parameter H/L.

From the second order Stokes theory the most important results concern:

(1) The form of the free surface:

$$\eta = \eta^{(1)} + \eta^{(2)} = \frac{H}{2}\cos(kx - \omega t) + \frac{kH^2}{16}\{3\coth^3(kh) - \coth(kh)\} \times \cos 2(kx - \omega t) \tag{1.13}$$

From a comparison of the wave forms of the two theories, depicted in Fig. 1.3, it is evident that the wave troughs develop flatter and the crests sharper according to 2nd order theory than according to 1st order theory. This is in closer agreement with observations of finite height waves which occur in nature.

(2) Horizontal velocity component function:

$$u = u^{(1)} + u^{(2)} = u^{(1)} + \frac{3c(kH)^2}{16} \cdot \frac{\cosh 2k(z+h)}{\sinh^4 kh} \times \cos 2(kx - \omega t) - \frac{gH^2}{8ch} \tag{1.14}$$

In addition to the periodic terms, a constant term appears, showing that finite amplitude waves transport mass as well as energy in the direction of their propagation. This is known as Stokes drift.

(3) Pressure distribution: the hydrodynamic pressure distribution given by the first order theory in Equation (1.7) is completed by a

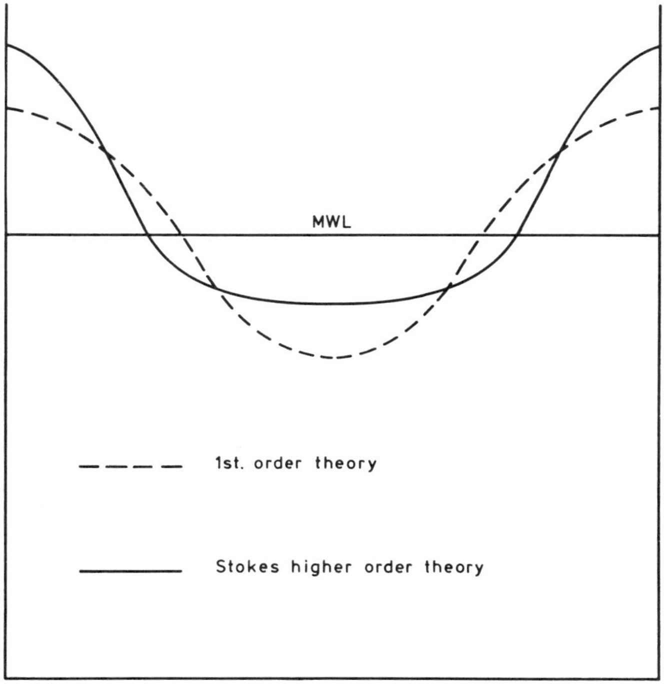

Fig. 1.3 Schematisation of the difference of the waveform given by 1st and 2nd order theories

second order term $p^{(2)}$,

$$p_{\text{dynamic}} = p^{(1)} + p^{(2)} = p^{(1)} + \frac{\rho g k H^2}{8 \sinh 2kh} \times \left[\left(\frac{3 \cosh 2k(z+h)}{\sinh^2 kh} - 1\right) \times \cos 2(kx - \omega t) - \cosh 2k(z+h) + 1\right] \tag{1.15}$$

where $p^{(1)}$ is the hydrodynamic pressure according to the first order theory.

1.4.2 Long waves

Close to the shore, the ratio $\varepsilon = h/L$ becomes much less than unity. If the inequality $HL/h^2 < 20 - 30$ is still valid and the boundary layer effect can be considered negligible, the velocity potential for irrotational flow is still applicable. The potential φ is developed in polynomial series with respect to the z variable,

$$\varphi = \varphi^{(0)}(x,t) + \varphi^{(1)}(x,t)\cdot(z+h) + \varphi^{(2)}(x,t)\cdot(z+h)^2 + \cdots \quad (1.16)$$

The retention of the $\varphi^{(0)}$ term in the mathematical model of Section 1.3 leads to the linear long wave equation:

$$\frac{\partial^2\eta}{\partial t^2} = c_0^2\frac{\partial^2\eta}{\partial x^2} \qquad \text{where } c_0 = \sqrt{(gh)} \quad (1.17)$$

If the first order term $\varphi^{(1)}(z+h)$ is retained, then, depending on the relation between $\delta = H/h$ and $\varepsilon = h/L$, three cases can be distinguished:

(1) $\delta \ll \varepsilon^2$
(2) $\delta \gg \varepsilon^2$
(3) $\delta = O[\varepsilon^2]$

(1) $\delta < \varepsilon^2$ or $UR = HL^2/h^3 \ll 1$. The differential equations in the surface velocity u and η can be linearised to the form:

$$\frac{\partial\eta}{\partial t} + h\frac{\partial u}{\partial x} = 0 \quad (1.18)$$

$$\frac{\partial u}{\partial t} + g\frac{\partial\eta}{\partial x} = \tfrac{1}{3}h^2\frac{\partial^3 u}{\partial x^2\,\partial t} \quad (1.19)$$

(2) $\delta > \varepsilon^2$ or $UR = HL^2/h^3 \gg 1$. The differential equations in u and η take the form:

$$\frac{\partial\eta}{\partial t} + \frac{\partial}{\partial x}[(h+\eta)\cdot u] = 0 \quad (1.20)$$

$$\frac{\partial u}{\partial t} + u\frac{\partial u}{\partial x} + g\frac{\partial\eta}{\partial x} = 0 \quad (1.21)$$

These are very useful non-linear long-wave equations. They are not frequency dispersive but amplitude dispersive, as the wave velocity is found to be $c = \sqrt{\{g(h+\eta)\}}$.

(3) $O[\delta] = O[\varepsilon^2]$ or $O[UR] = 1$. The equations take the form, known as Boussinesq equations,

$$\frac{\partial\eta}{\partial t} + \frac{\partial}{\partial x}[(h+\eta)u] + \tfrac{1}{3}h^3\frac{\partial^3\eta}{\partial x^3} = 0 \quad (1.22)$$

$$\frac{\partial u}{\partial t} + u\frac{\partial u}{\partial x} + g\frac{\partial\eta}{\partial x} = 0 \quad (1.23)$$

The special case of long waves progressing in one direction is described by the Korteweg–De Vries equation

$$\frac{1}{\sqrt{(gh)}}\frac{\partial\eta}{\partial t}+\frac{\partial\eta}{\partial x}+\frac{3}{2h}\eta\frac{\partial\eta}{\partial x}+\tfrac{1}{6}h^2\frac{\partial^3\eta}{\partial x^3}=0 \tag{1.24}$$

The solution of (1.24) for specific boundary conditions, is the solitary wave with infinite period. The general solution of (1.24), after the observation that η is a function of the θ parameter, where

$$\frac{\theta}{2k}=\frac{t}{T}-\frac{x}{L} \tag{1.25}$$

is the cnoidal wave, approximating, in an optimal way, the waves in shallow water before breaking.

1.5 NUMERICAL SOLUTION OF LONG-WAVE MODEL

Numerical solutions for the short wave (Boussinesq) models have been successfully investigated and computer packages produced in European and US research centers. Their presentation are beyond the introductory scope of this book and we shall limit ourselves to the simpler long-wave models.

Long waves with length at least an order of magnitude greater than the depth and wave height a finite proportion of the depth have a very important place in maritime hydraulics. Wind generated waves near the coast, tides, free and forced oscillation in coastal basins (natural or artificial) and tsunamis are dominated and described by the long-wave model (in linear or non-linear form). Some simple, widely used algorithms for their numerical solution together with some illustrative applications are within the scope of this section.

In x, t space the field equations, deriving from equilibrium of forces and mass conservation principles, are respectively,

$$\frac{\partial u}{\partial t}+u\frac{\partial u}{\partial x}+g\frac{\partial\eta}{\partial x}=0 \quad\text{or}\quad =\frac{-\tau_b}{\rho(h+\eta)} \tag{1.26}$$

$$\frac{\partial}{\partial t}(h+\eta)+\frac{\partial}{\partial x}[(h+\eta)u]=0 \quad\text{or}\quad \frac{\partial\eta}{\partial t}+\frac{\partial}{\partial x}[(h+\eta)u]=0 \tag{1.27}$$

The unknown functions to be described as solutions of (1.26), (1.27) are $u(x,t)$ and $\eta(x,t)$, while the initial depth $h(x)$ is considered known. The assumption of real viscous fluid can introduce in (1.26) a frictional loss term. If the bed shear is expressed as a quadratic form of the particle velocity,

$$\frac{\tau_b}{\rho} = k \cdot |u| u \tag{1.28}$$

where k is a nondimensional bed friction coefficient, (1.26) takes the modified form. The gravity wave described by (1.26), (1.27) propagates in the positive and negative x directions with velocity c, given by:

$$c = \sqrt{\{g(h + \eta)\}} \tag{1.29}$$

This is an amplitude dispersive wave and the wave crests propagate at higher speed than the wave troughs. The result is the continuous deformation of the wave up to the point of breaking (when the downward slope of the wave exceeds some critical value). Dropping the non-linear terms in (1.26), (1.27) leads to a simpler form:

$$\frac{\partial \eta}{\partial t} + h \frac{\partial u}{\partial x} = 0 \tag{1.30}$$

$$\frac{\partial u}{\partial t} + g \frac{\partial \eta}{\partial x} = 0 \tag{1.31}$$

The elimination of the u function from the pair (1.30), (1.31) leads to one second order hyperbolic equation in the $\eta(x, t)$ function (Equation (1.17)). The linear Equation (1.17) describes the propagation without deformation (for $h = cte$) of pulses in both x^+ and x^-, at speed $c = \sqrt{(gh)}$.

The solution of (1.26), (1.27) or (1.30), (1.31) can follow various paths. The application of the method of characteristics can give a sequence of u and $c = \sqrt{(gh)}$ values on the characteristic curves or in the x, t plane. This method is, from the physical point of view, the most accurate as it can follow the development of the wave form, even after the generation of shocks (wave breaking), without numerical errors. As the analysis and programming is rather complicated various finite difference and finite element schemes have been developed for a direct description of the evolution of waves in the x, t plane.

Some simple finite difference schemes are presented here. Their performance for the non-linear model is limited, but they behave well in the case of the linearised model. Three different phenomena are analysed:

(1) The propagation of linear and non-linear waves in an infinite field of constant depth.

(2) The propagation of linear waves with shoaling over a sloped bed.

(3) The partial reflection of linear waves over a reef (submerged breakwater).

The boundary conditions that complete the two field equations are the following:

(1) Free transmission boundary without back reflection. This can be a natural absorbing boundary, or an artificial boundary, imposed for computational reasons in order to keep finite the field within which our computational interest lies. The condition on this boundary derives from conservation principles in the propagation of a moving surge. It relates the free surface elevation to the particle velocity,

$$\mathbf{n} \cdot \mathbf{u} = \frac{n}{h} \sqrt{(gh)} \tag{1.32}$$

where **n** is the outward normal unit vector.

(2) Incidence boundary, where incoming waves are known and prescribed and any back reflected wave is permitted to cross that boundary. On that boundary the free surface elevation function η is analysed in two parts, an incident η_i and a radiated η_r. The second is dominated by the differential equation describing the propagation of a disturbance in one direction (namely the outward) with speed $= c$,

$$\frac{\partial \eta_r}{\partial t} + \sqrt{(gh)} \frac{\partial \eta_r}{\partial n} = 0 \tag{1.33}$$

(3) Reflection boundary. On this boundary the normal velocity is suppressed, resulting in the suppression of the normal derivative of η ($\partial \eta / \partial n = 0$).

1.5.1 Propagation of linear and non-linear waves in constant depth

The numerical solution is performed by explicit finite difference schemes on a staggered grid. Due to the grid form, u and η are computed at different locations in the x, t plane. The space and time discretisation is schematically given in Fig. 1.4.

Forward differences is used for the time derivative and centered differences in space. The continuity and equilibrium equations, according to the notation of Fig. 1.4, take the form:

$$\frac{\eta_i^{n+1} - \eta_i^n}{\Delta t} = -\frac{(h_i^n + h_{i+1}^n)u_{i+1}^n - (h_i^n + h_{i-1}^n)u_i^n}{2\Delta x} \tag{1.34}$$

$$\frac{u_i^{n+1} - u_i^n}{\Delta t} = -\frac{(u_{i+1}^n + u_i^n)^2 - (u_i^n + u_{i-1}^n)^2}{8\Delta x} - \frac{g(\eta_i^n - \eta_{i-1}^n)}{\Delta x} \tag{1.35}$$

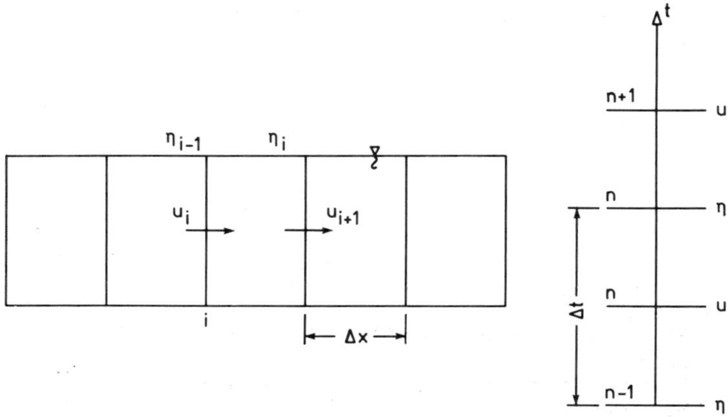

Fig. 1.4 Spatial and temporal discretisation for 1-D long-wave model

Other types of finite differences for the convective term can be used as alternatives to that in (1.35):

$$u\frac{\partial u}{\partial x} \simeq \frac{u_i^n(u_{i+1}^n - u_{i-1}^n)}{2\Delta x}$$

$$\text{or} \qquad u_i^n\frac{-u_{i+2}^n + 8u_{i+1}^n - 8u_{i-1}^n + u_{i-2}^n}{12\Delta x} \tag{1.36}$$

Backward differences for the space derivatives, which are stable, can be used instead of the centered differences. The model takes the form:

$$\frac{\eta_i^{n+1} - \eta_i^n}{\Delta t} = -u_i^n\frac{(h_i^n - h_{i-1}^n)}{\Delta x} - h_i^n\frac{(u_{i+1}^n - u_i^n)}{\Delta x} \tag{1.37}$$

$$\frac{u_i^{n+1} - u_i^n}{\Delta t} = -u_i^n\frac{(u_i^n - u_{i-1}^n)}{\Delta x} - \frac{g}{\Delta x}(\eta_i^n - \eta_{i-1}^n) \tag{1.38}$$

The numerical stability of the explicit schemes presented above is ensured for values of the Courant number,

$$(u + \sqrt{(gh)})\frac{\Delta t}{\Delta x} < 1 \tag{1.39}$$

A computer program in BASIC synthesized on the basis of the centered scheme, and refering to a flow domain of variable depth with an incoming wave (from the left boundary) of given period and incident height H_i, and an outgoing wave from the right boundary with no reflection, is given here. Program 1 is used with appropriate modification for the description of the propagation of linear waves in Sections 1.5.2 and 1.5.3.

PROGRAM 1: NON-LINEAR 1-D LONG WAVE MODEL FOR PROPAGATION OVER CONSTANT DEPTH

```
   10REM LONG NONLINEAR 1-D WAVE MODEL PROPAGATION
 OVER CONSTANT DEPTH CENTERED DIFFERENCES STAGGERE
D GRID
   20DIM U(51),UN(51),H(51),HN(51),Z(51),HO(51)
   30READ DT,DX,IM,NM,ZO,PER,HO
   40DATA...
   50FORI=OTOIM:READHO(I):H(I)=HO(I):NEXTI
   55DATA...
   60N=0:T=0:C=SQR(9.81*H(1)):L=C*PER
   70N=N+1:T=T+DT
   80IF (N-1)*DT>DX/C THEN GOTO100
   90Z1=0:GOTO 130
  100Z1=Z(1)-ZO*SIN(2*PI*(N-1)*DT/PER)
  110Z2=Z(2)-ZO*SIN(2*PI*(N-1)*DT/PER-DX/L)
  120Z1=Z1+DT/DX*C*(Z2-Z1)
  130Z(1)=ZO*SIN(2*PI*N*DT/PER)+Z1
  140HN(1)=HO(1)+Z(1)
  150FORI=2TOIM-1:HN(I)=H(I)-DT/DX/2*((H(I)+H(I+1)
)*U(I+1)-(H(I)+H(I-1))*U(I)):NEXTI
  160FORI=2TOIM-1:Z(I)=HN(I)-HO(I):NEXTI
  170FORI=2TOIM-1:UN(I)=U(I)-DT/DX/8*((U(I+1)+U(I)
)^2-(U(I)+U(I-1))^2)-9.81*DT/DX*(Z(I)-Z(I-1)):NEXT
I
  180UN(1)=UN(2):UN(IM)=Z(IM-1)*SQR(9.81/HN(IM-1))

  190FORI=1TOIM:H(I)=HN(I):U(I)=UN(I):NEXTI
  200IFN/20<>INT(N/20) THEN GOTO70
  210PRINTT
  220FORI=1TOIM:PRINTZ(I);:NEXTI:PRINT
  230FORI=1TOIM:PRINTU(I);:NEXTI
  240IFN<NM THEN GOTO70
  250END
>
```

Description of main variables:

DT, DX = time and space discretisation steps
IM, NM = max values of space (I) and time (N) steps
Z0, PER = amplitude and period of incoming waves
Z = free surface elevation with respect to the SWL
H0 = initial water depths
H, HN = past and present water depths
U, UN = past and present velocity values

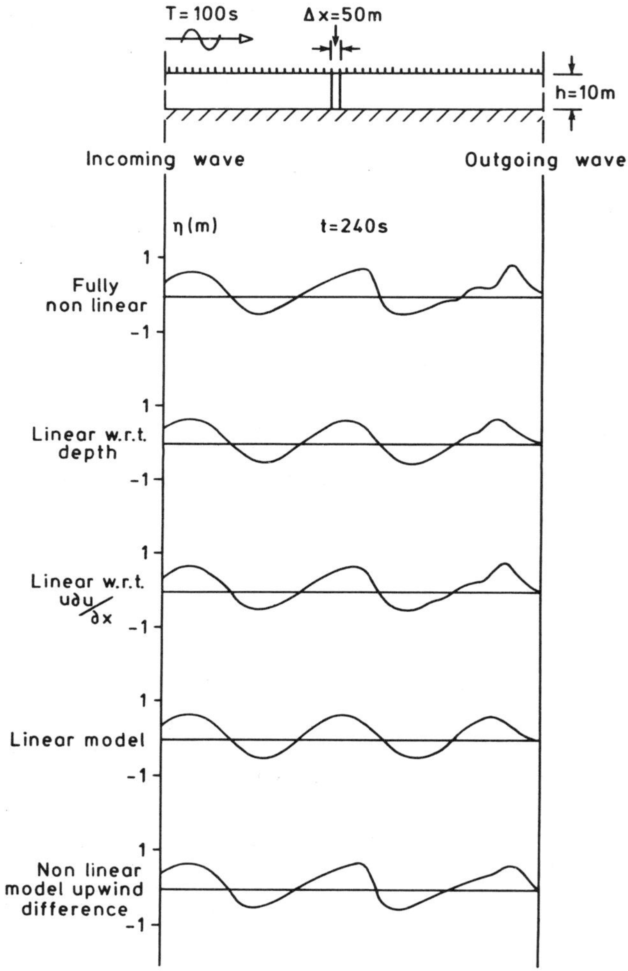

Fig. 1.5 Flow domain form and discretisation, wave form at $t = 240$ s for various numerical schemes

The application refers to the flow domain depicted and discretised in Fig. 1.5. The object is to present the evolution and deformation of an initially sinusoidal wave to a form containing higher frequencies. The wave form after 240 s from the initiation of the flow for the non-linear, the fully linear and the linearised (with respect to the convective term and to the water depth) models is depicted in Fig. 1.5. It is evident that the numerical scheme presented is frequency dispersive (numerical dispersion error). The same behaviour is shown by the upwind (backward) differences. The data used were: DT = 2 s, DX = 50 m, IM = 51, NM = 500, Z0 = 1 m, PER = 100 s, depth ≡ 10 m.

1.5.2 Propagation with shoaling over a sloping bed

Although the analytical solution for the evolution of the wave height, based on energy conservation, is simple (given by Equation (1.44)) the numerical solution is presented and compared with the analytical one.

The flow domain has a depth varying from 10 m to 3 m. Beyond that depth coastwards, breaking of the wave and absorption of its energy is not described by the model, but it is simulated through the use of the free radiation boundary condition on the right boundary (Fig. 1.6).

The conservation of wave power between two sections

$$P_1 = \frac{\gamma H_1^2 L_1}{8} = P_2 = \frac{\gamma H_2^2 L_2}{8} \tag{1.40}$$

in the case of long waves ($L = cT = \sqrt{(gh)}T$) leads to the following relation between the depth and the wave height:

$$\frac{L_1}{L_2} = \sqrt{\left(\frac{h_1}{h_2}\right)}, \qquad \frac{H_1}{H_2} = \sqrt[4]{\left(\frac{h_2}{h_1}\right)} \tag{1.41}$$

The numerical solution was done by the explicit centered finite difference scheme described in Section 1.5.1 for $\Delta x = 20$ m, $\Delta t = 1$ s. The free surface form, 110 s after the initiation of the wave propagation, is depicted in Fig. 1.6. The increase in height with the decrease in depth is apparent. Table 1.2 gives a comparison between the analytically and numerically found values of H and L for various depths.

1.5.3 Partial reflection of long waves over a shoal (submerged breakwater)

The sudden decrease in water depth, in the case of a progressive wave, causes partial reflection of the wave, and transmission of the wave beyond the shoal with decreased height. This results in protection of the coast from high wave energy by natural or artificial means. If progress over the shoal (or submerged breakwater) does not involve wave breaking, then the wave energy conservation principle implies the following relation for the incident H_i, reflected H_r and transmitted

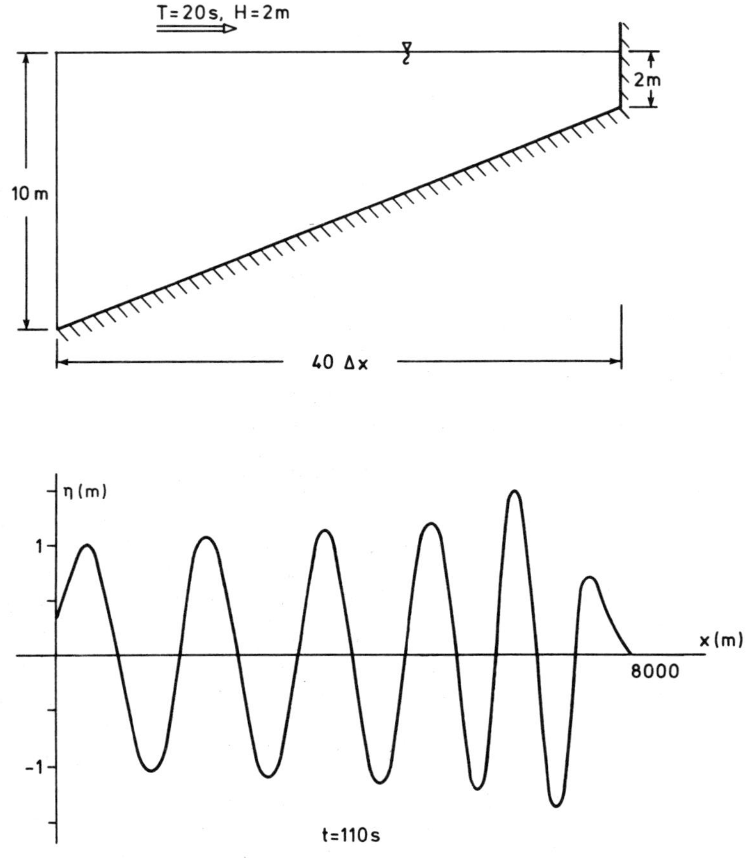

Fig. 1.6 Long-waves shoaling over a sloping bed

Table 1.2

h (m)	10	8	7	4.5	3
H analyt.	2	2.12	2.27	2.45	2.70
H numer.	2	2.2	2.4	2.55	2.75
h (m)	9	7	5.5	4	3
L analyt.	186	164	146	124	107
L numer.	160	155	144	124	104

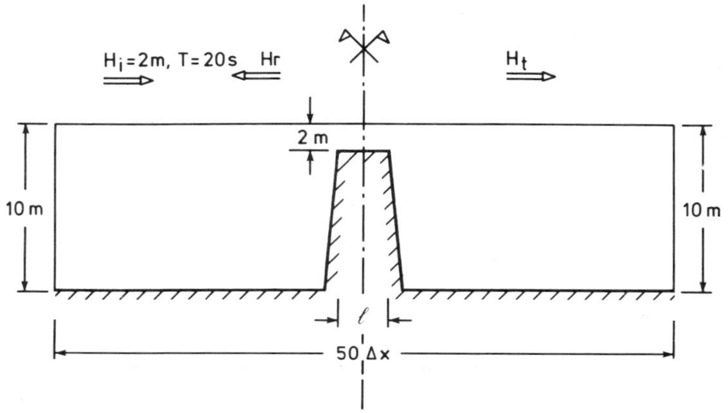

Fig. 1.7 Submerged obstacle—flow domain morphology

H_t wave heights:

$$H_i^2 = H_r^2 + H_t^2 \tag{1.42}$$

It is obvious that the H_r, H_t magnitudes are functions of h_1, h_2 (water depth before and above the shoal) and l, L (shoal length and wave length). The phenomenon has been extensively investigated in the laboratory and analytically. The numerical solution given here is the only mathematical approach for more complex situations (variable water depth, successive shoals, non-periodic waves, etc.).

The application was done for 'before' and 'on' shoal depths of 10 m and 2 m, respectively, wave period 20 s and ratios of shoal length to wave length of 0.3, 0.4, 0.6 and 0.9. The phenomenon is shown schematically in Fig. 1.7. The upstream boundary condition is of the *known incident wave and free radiation of reflected wave type*; the downstream one is of the *free radiation type*.

The linear long wave model was solved for $\Delta t = 4$ s, $\Delta x = 40$ m, $T = 20$ s and $l = 1, 2, 3, 5\Delta x$; the free surface profile along the flow domain for $l = 2\Delta x$ at a certain time instant is given in Fig. 1.8.

From the computed transmitted wave heights the diagram of Fig. 1.9 was synthesized depicting the influence of the shoal length on the wave for a given ratio, h_1/h_2. It seems that the shoal influence becomes significant for $l/L = 0.3$ and the transmitted wave becomes negligible for $l/L > 0.6$.

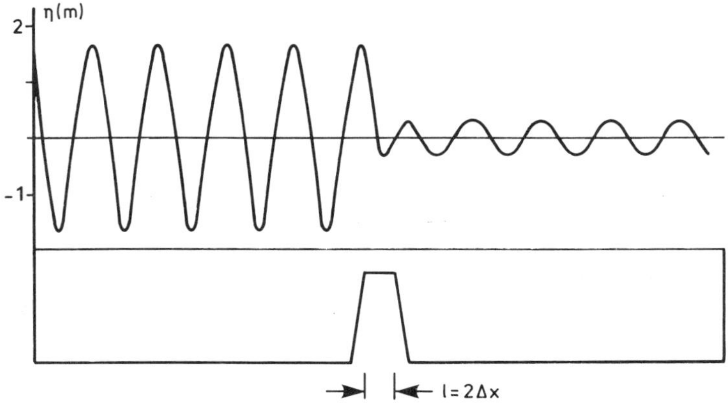

Fig. 1.8 Waveform for given time showing reflected, incident and transmitted waves

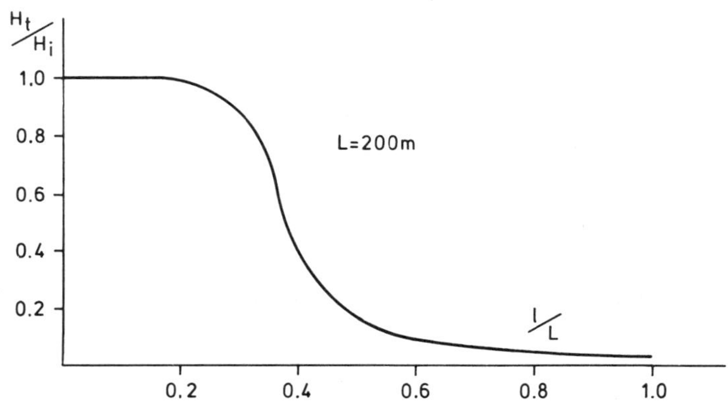

Fig. 1.9 Wave transmission as a function of relative length of obstacle for given height

1.6 WAVE DEFORMATION NEAR THE COAST

Careful observation of Equations (1.8) for frequency dispersion and (1.10) for energy conservation reveal that even for an ideal fluid the wave elements (H, L) are transformed when the wave propagates to areas of different depths. The phenomena which occur in the near coastal region and affect the wave elements are as follows:

(1) Shoaling (influence of depth decrease)
(2) Refraction

(3) Breaking
(4) Energy dissipation (due to bed friction).

1.6.1 Shoaling effect

Under the assumptions of (1) two dimensionality (long crested waves), (2) no frictional losses, (3) wave period conservation, Equation (1.10), for wave power applied at two cross-sections of different depth, results in:

$$P_1 = P_2 \rightarrow \frac{\gamma H_1^2 L_1 n_1}{8} = \frac{\gamma H_2^2 L_2 n_2}{8}$$

$$\rightarrow k_s = \frac{H_2}{H_1} = \sqrt{\left(\frac{n_1 L_1}{n_2 L_2}\right)} \tag{1.43}$$

The computational procedure for the estimation of k_s (shoaling coefficient) starts with the determination of the wave length L from the wave period and depth by use of Equation (1.86), and the determination of the wave parameter n by use of (1.10). If the infinite depth h_∞ and the corresponding $H_0, L_0 = gT^2/2\pi, n_0 = 1/2$ are used as reference magnitudes, (1.43) takes the form,

$$H = H_0 \sqrt{\left(\frac{L_0}{2nL}\right)} = H_0 \cdot k_s \tag{1.44}$$

Although the functions $n, L/L_0, k_s$ have been tabulated and are available in most coastal engineering textbooks, the computational procedure in BASIC is presented here (Program 2), as a routine to be incorporated in more complex programs of wave models.

PROGRAM 2: WAVE SHOALING

```
>L.
   10REM WAVE SHOALING
   20READ T,H,HO
   30DATA...
   40LO=9.81*T^2/2/PI
   50MN=LO:IT=0:L=LO
   60IT=IT+1:IF IT>100 THEN GOTO150
   70A=EXP(2*PI*H/MN):B=EXP(-2*PI*H/MN)
   80TP=L:L=LO*(A-B)/(A+B)
   90IFABS(TP-L)<.0001 THEN GOTO110
  100MN=(TP+L)/2:GOTO60
  110A=4*PI*H/L
  120N=.5*(1+A/(EXP(A)-EXP(-A))*2)
  130H1=HO*SQR(.5*LO/N/L)
  140PRINTN,H1,L
  150END
>
```

Description of variables:

H0 = initial wave height for infinite depth
T = wave period
H = water depth where the wave elements are to be computed
L, H1 = wave length and wave height at depth H.

The procedure is simply the solution of a trigonometric equation with successive iterations.

1.6.2 Wave refraction in water of intermediate depth

The dispersion relation (1.8) reveals that the wave celerity depends both on the wave period and water depth. Consequently waves propagating in water of varying depth may suffer refraction, analogous to refraction of light and sound. Such refraction results in change of the direction of wave propagation and to 3-dimensionalisation of a 2-dimensional long-crested wave.

Snell's law of optic refraction, using the notation of Fig. 1.10, is

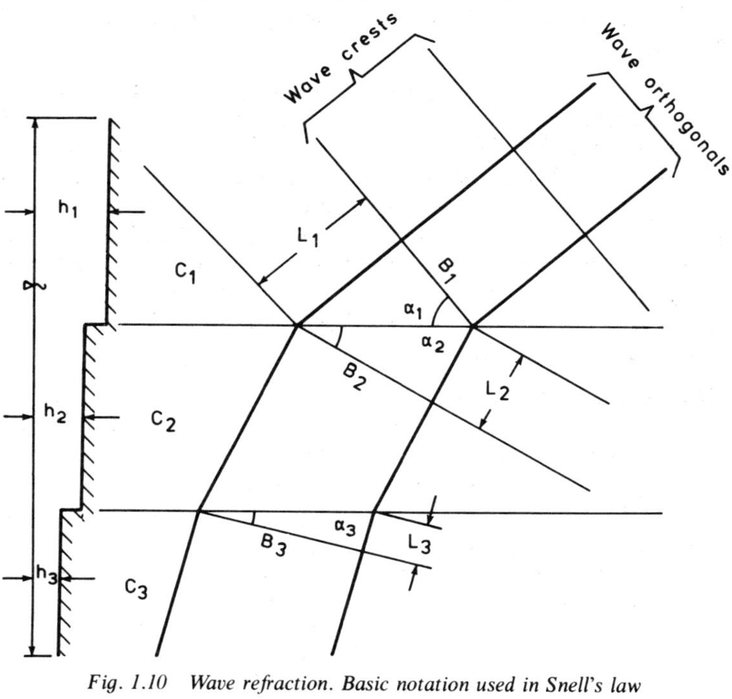

Fig. 1.10 Wave refraction. Basic notation used in Snell's law

written as

$$\frac{L_1}{L_2} = \frac{\sin a_1}{\sin a_2} \rightarrow a_2 = \arcsin\left(\sin a_1 \cdot \frac{L_2}{L_1}\right) \tag{1.45}$$

$$\frac{B_2}{B_1} = \frac{\cos a_2}{\cos a_1} \rightarrow B_2 = B_1 \frac{\cos a_2}{\cos a_1} \tag{1.46}$$

The relation of the two wave heights corresponding to the regions #1 and #2 is based on the wave power propagation in the stream tube confined between two orthogonal waves:

$$B_1 \frac{\gamma H_1^2 L_1 n_1}{8} = B_2 \frac{\gamma H_2^2 L_2 n_2}{8} \tag{1.47}$$

giving

$$\frac{H_2}{H_1} = \sqrt{\left(\frac{L_1 n_1}{L_2 n_2}\right)} \sqrt{\left(\frac{B_1}{B_2}\right)} = k_s \cdot k_r \tag{1.48}$$

where k_r is the refraction coefficient.

From a computational point of view, the solution of a refraction problem leads to the determination of the wave direction and height inside a certain coastal domain. Two procedures are currently followed:

(1) The first aims at the graphical or computational construction of a number of wave orthogonals (and subsequently to the determination of the k_r coefficients from the distance between orthogonals). The construction is based on relations (1.45), (1.46) and geometrical and kinematic principles.

(2) The second aims at the determination of θ and E (the wave direction angle with respect to a chosen coordinates system and the wave energy giving the wave height H) at all computation points inside the flow domain. The method is based on the wave energy and wave number conservation equations.

According to the first method of constructing wave orthogonals (in its computational form) the orthogonals are approximated piecewise linearly and the sequence of x_i, y_i pairs of the coordinates of the apexes are computed. We begin by describing the evolution of a wave orthogonal along a length increment ds (or a time increment $\mathrm{d}t = \mathrm{d}s/c$) according to Fig. 1.11.

Using kinematic principles, the following equalities are formulated:

$$\frac{\mathrm{d}x}{\mathrm{d}t} = c \cdot \cos\theta \tag{1.49}$$

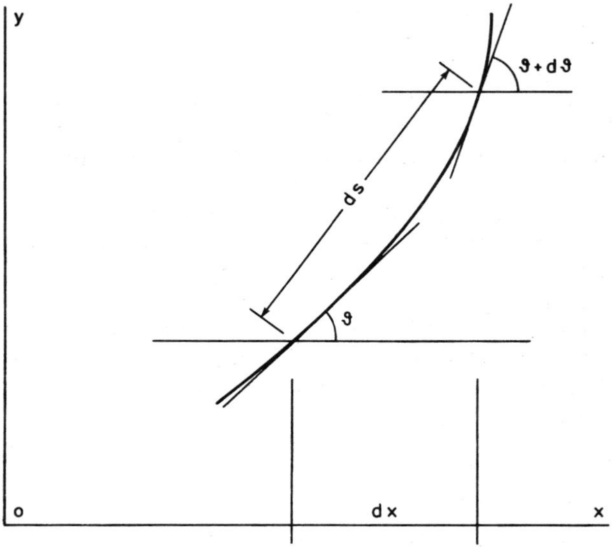

Fig. 1.11 Wave refraction. Geometry of a bent orthogonal

$$\frac{dy}{dt} = c \cdot \sin\theta \tag{1.50}$$

$$\frac{d\theta}{dt} = \frac{\partial c}{\partial x}\sin\theta - \frac{\partial c}{\partial y}\cos\theta \tag{1.51}$$

$$\frac{d\theta}{ds} = \frac{1}{c}\left(\frac{\partial c}{\partial x}\sin\theta - \frac{\partial c}{\partial y}\cos\theta\right) \tag{1.52}$$

Equations (1.49) to (1.51) are used for the computation of the successive coordinate pairs (x_i, y_i) and the corresponding slope θ_i of the line segment approximating locally the wave orthogonal.

From wave power conservation applied to a wave crest, the following equation is formulated for computation of the distance between the orthogonals β (and the refraction coefficient).

$$\frac{d^2\beta}{dt^2} + p_t\frac{d\beta}{dt} + q_t\beta = 0 \tag{1.52}$$

where

$$p_t = -2\left(\frac{\partial c}{\partial x}\cos\theta + \frac{\partial c}{\partial y}\sin\theta\right) \tag{1.53}$$

$$q_t = c\left(\frac{\partial^2 c}{\partial x^2}\sin^2\theta - \frac{\partial^2 c}{\partial x\,\partial y}2\sin\theta\cos\theta + \frac{\partial^2 c}{\partial y^2}\cos^2\theta\right) \quad (1.54)$$

An algorithm is now presented for the computation of a wave orthogonal based on Equations (1.49)–(1.51). Its repetition for a series of orthogonals with a given initial distance and direction in the open sea (depth $> L/2$) gives an invaluable insight into the evolution of waves in a coastal area. The graphical representation (plotting) of the set of orthogonals can be done using the 'graphics' statements available in most contemporary microcomputers.

The numerical treatment of (1.49)–(1.51) starts with the discretisation of the flow domain by means of an orthogonal grid of variable step, DX, DY. The wave velocity c is computed at the nodes of the grid from the known depth. The spatial derivatives $\partial c/\partial x$, $\partial c/\partial y$ are approximated by centered finite differences,

$$\frac{\partial c}{\partial x} = \frac{c_{i+1j} - c_{i-1j}}{\Delta x_i + \Delta x_{i-1}}, \qquad \frac{\partial c}{\partial y} = \frac{c_{ij+1} - c_{ij-1}}{\Delta y_j + \Delta y_{j-1}} \quad (1.55)$$

Starting with the coordinates, x^0, y^0 and the direction θ^0 of a wave orthogonal in the open sea as boundary conditions, the numerical solution of (1.49)–(1.51) follows an explicit (shooting) procedure and leads to the computation of sets x^n, y^n, θ^n. Equation (1.51) is solved for the new θ value

$$\theta^{n+1} - \theta^n = \Delta\theta = \Delta s \cdot k_s = \Delta s\left(\frac{\sin\theta^n}{c^n}\cdot\frac{\partial c^n}{\partial x} - \frac{\cos\theta^n}{c^n}\cdot\frac{\partial c^n}{\partial y}\right) \quad (1.56)$$

where

$$\Delta s = \Delta t \cdot c^n \quad (1.57)$$

The computation of the θ^{n+1} value is followed by the computation of x^{n+1}, y^{n+1} by means of (1.49), (1.50)

$$x^{n+1} = x^n + \Delta s \cdot \cos\left(\frac{\theta^n + \theta^{n+1}}{2}\right) \quad (1.58)$$

$$y^{n+1} = y^n + \Delta s \cdot \sin\left(\frac{\theta^n + \theta^{n+1}}{2}\right) \quad (1.59)$$

As the $x^n, y^n, x^{n+1}, y^{n+1}$ points do not in general coincide with grid nodes it is necessary to interpolate values of c, $\partial c/\partial x$, $\partial c/\partial y$ appearing in (1.56), (1.57) from known nodal values of those magnitudes.

According to the notation of Fig. 1.12, the following interpolation formula is used:

$$c^n = c_{ij}(\xi - 1)(\eta - 1) - c_{i+1j}(\eta - 1)\xi + c_{i+1j+1}\xi\eta + c_{ij+1}\eta(\xi - 1) \quad (1.60)$$

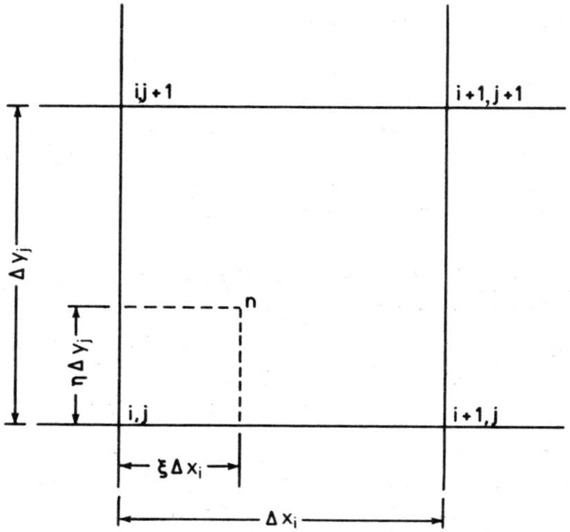

Fig. 1.12 Four-point interpolation used in wave refraction model

When the computation of the $x^0y^0, x^1y^1 \cdots x^ny^n \cdots$, sequence describing the orthogonal from the open sea to the coast (where the depth goes to zero) is completed, the results are printed or plotted and the computation goes on to the next orthogonal. The procedure is programmed in BASIC (Program 3) and the graphics commands correspond to that of a BBC micro, easily adaptable to any other type of micro.

PROGRAM 3: WAVE REFRACTION MODEL ORTHOGONALS COMPUTATION

```
>LIST
    5REM WAVE REFFRACTION MODEL ORTHOGONALS COMPUT
ATAION
   10DIMC(20,20),D(20,20),XO(20),YO(20),X(300),Y(3
00),C1(50),CX(20,20),CY(20,20),DX(20),DY(20)
   20READ T,IM,JM,NM,SM,DT,HO,DM
   30DATA...
   40FORJ=1TOJM:FORI=1TOIM:READD(I,J):NEXTI:NEXTJ
   50DATA...
   60FORI=1TOIM-1:READDX(I):NEXTI
   70DATA...
   80FORJ=1TOJM-1:READDY(J):NEXTJ
   90DATA...
  100SUMX=0:FORI=1TOIM-1:SUMX=SUMX+DX(I):NEXTI:SUM
```

```
Y=0:FORJ=1TOJM-1:SUMY=SUMY+DY(J):NEXTJ
  105FORK=1TODM
  110L0=9.81*T^2/2/PI:C0=9.8*T/2/PI
  120L=L0:TM=L
  130MN=(L+TM)/2:TM=L:A=2*PI*K/MN:L=L0*(EXP(A)-EXP
(-A))/(EXP(A)+EXP(-A))
  140IF(ABS(L-TM)>.1)THEN GOTO130
  150C1(K)=C0*(EXP(A)-EXP(-A))/(EXP(A)+EXP(-A))
  160NEXTK
  170FORJ=1TOJM:FORI=1TOIM:IFD(I,J)=0 THEN GOTO200
  180FORK=1TODM
  182IFD(I,J)<K THEN GOTO190
  184NEXTK
  190C(I,J)=C1(K-1)+(C1(K)-C1(K-1))*(K-D(I,J))

  200NEXTI :NEXTJ:CLS
  210XX=0:FORI=1TOIM-1:XX=XX+DX(I):XXX=XX*1000/SUM
X:MOVE XXX,0:DRAW XXX,1000:NEXTI
  220YY=0:FORJ=1TOJM-1:YY=YY+DY(J):YYY=YY*1000/SUM
Y:MOVE 0,YYY:DRAW 1000,YYY:NEXTJ
  225FORI=1TOIM-1:FORJ=1TOJM-1
  230CX(I,J)=(C(I+1,J)+C(I+1,J+1)-C(I,J)-C(I,J+1))
/2/DX(I)
  240CY(I,J)=(C(I,J+1)+C(I+1,J+1)-C(I,J)-C(I+1,J))
/2/DY(J)
  250NEXTJ:NEXTI
  260FORN=1TONM:READX0(N),Y0(N):NEXTN
  270DATA...
  280FORN=1TONM :H=RAD(H0):X(1)=X0(N):Y(1)=Y0(N):F
ORST=2TOSM
  290XX=0 :FORI=1TOIM:XX=XX+DX(I)
  292IFXX>X(ST-1)THENGOTO 300
  294NEXTI
  300XX=(X(ST-1)-XX+DX(I))/DX(I)
  310YY=0:FORJ=1TOJM :YY=YY+DY(J)
  312IFYY>Y(ST-1) THEN GOTO320
  314NEXTJ
  320YY=(Y(ST-1)-YY+DY(J))/DY(J)
  330A1=(XX-1)*(YY-1):A2=-XX*(YY-1):A3=XX*YY:A4=-Y
Y*(XX-1)
  340CC=C(I,J)*A1+C(I+1,J)*A2+C(I+1,J+1)*A3+C(I,J+
1)*A4 :IF(CC=0)THEN ST=ST-1:GOTO430
  350CX=CX(I,J)*A1+CX(I+1,J)*A2+CX(I+1,J+1)*A3+CX(
I,J+1)*A4
  360CY=CY(I,J)*A1+CY(I+1,J)*A2+CY(I+1,J+1)*A3+CY(
I,J+1)*A4
  370KS=SIN(H)/CC*CX-COS(H)/CC*CY:DS=DT*CC:DH=KS*D
S:HM=H+DH/2
  380X(ST)=X(ST-1)+DS*COS(HM)
  390Y(ST)=Y(ST-1)+DS*SIN(HM)
```

```
  400H=H+DH
  410IF X(ST)<0 OR X(ST)>SUMX OR Y(ST)<0 OR Y(ST)>
SUMY THEN ST=ST-1:GOTO430
  420NEXTST
  430MOVEX(1)*1000/SUMX,Y(1)*1000/SUMY
  440FORSS=1TOST-1:DRAWX(SS)*1000/SUMX,Y(SS)*1000/
SUMY:NEXTSS
  450NEXTN
  460END
```

Description of variables:

C(I, J)	= wave velocity referring to the node I, J of the grid
D(I, J)	= corresponding water depth
X0(K), Y0(K)	= coordinates of kth orthogonal at the starting point (open sea condition)
X(ST), Y(ST)	= coordinates of the ST^{th} apex of an orthogonal
C1(N)	= wave velocity corresponding to a depth = N meters
IM, JM	= number of grid nodes along X, Y of the flow domain
NM	= number of computed orthogonals
SM	= maximum number of apexes of an orthogonal
DX, DY	= space discretisation steps along Ox, Oy
DT	= time discretisation step in the description of an orthogonal (DS = DT*C)
H0	= initial direction of the orthogonals (θ^0 referring to the open sea, and measured from the Ox axis)
DM	= maximum water depth in the studied coastal region in metres (integer)

The application presented here is intended to control the accuracy of the numerical procedure by comparison with the analytical solution existing in the case of parallel depth contours. The coast morphology is given in Fig. 1.13. Two orthogonals are described, characterised by $\theta^0 = 45°$. The wave has period $T = 10$ s. The rest of the program data are: IM = 6, JM = 6, NM = 2, SM = 10, DS = 100 m (ds is fixed in this case and not dt), H0 = 45°, DX = 400 m, DY = 200 m, X(0) = 400, Y(0) = 0, X(0) = 800, Y(0) = 0. The following coordinate pairs are found for the two orthogonals:

First:	400	469	536	600	661	717	770	917	858	877
	0	72	147	223	303	385	471	559	650	748
Second:	800	869	936	1000	1061	1117	1170	1217	1258	1277
	0	72	147	223	303	385	471	519	650	748

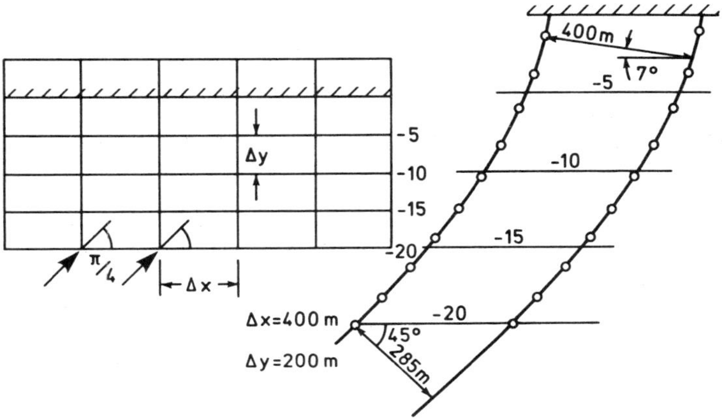

Fig. 1.13 Wave refraction. Flow domain and computed orthogonals

The graphical reproduction of the two orthogonals in Fig. 1.13 shows that at depth 2.5 m the angle between orthogonals and isodepths is 82° and that the distance between the two orthogonals changes from 285 m to 400 m. An analytical solution based on relations (1.45) and (1.46) leads to similar numbers.

A more sophisticated approach to the solution of a refraction problem resulting in the computation of both the wave angle and the refracted wave height (very useful for the direct computation of radiation stresses) is by making use of the modified form of the general model for refraction–diffraction. The model is formulated on the basis of wave number and wave energy conservation principles, mild bed slope, steady wave conditions and only depth refraction. It takes the following form:

$$\frac{\partial}{\partial x}\left(\frac{\sin\theta}{c}\right)=\frac{\partial}{\partial y}\left(\frac{\cos\theta}{c}\right) \tag{1.61}$$

$$\frac{\partial}{\partial x}(H^2 c_g \cos\theta)+\frac{\partial}{\partial y}(H^2 c_g \sin\theta)=0 \tag{1.62}$$

where the coordinates and the wave angle θ are oriented according to the notation of Fig. 1.14. These are two first order PDE's in the unknown variables $\theta(x, y)$ and $H(x, y)$; the wave velocity c and the group velocity $c_g = nc$ are known functions of the wave period T and the known local depth $h(x, y)$.

Various finite element and finite difference schemes have appeared recently for the solution of the refraction model (1.61) and (1.62). Here it is treated as an initial value problem. The wave height and

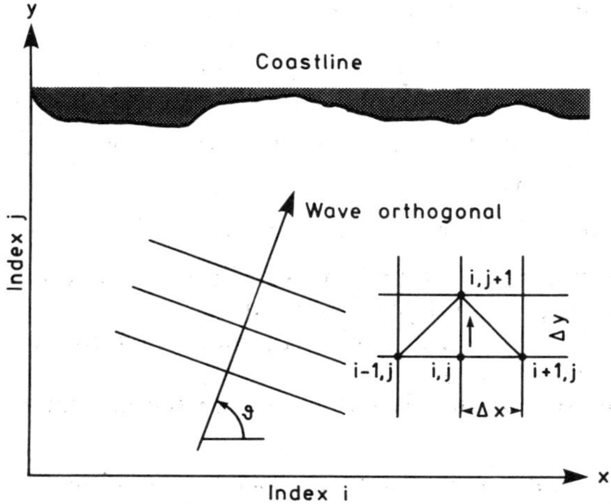

Fig. 1.14 Coordinate axes and notation used in refraction model

angle are assumed known on the open sea (deep water) boundary and the solution can proceed in the direction of the coast by successive computations of H and θ from previous known values in an explicit (shooting) type of operation. This procedure, depending on the Δx value and the wave celerity, can lead to a numerical stability problem which can be resolved by the proper selection of the Δy step along the direction of the propagation (1st order hyperbolic equation solved by upwind differences).

A further concern is the wave angle equation when the wave angle passes the $\theta = 0$ or $\theta = \pi$ region ($\sin\theta \to 0$). In such regions the sign of the increment $\delta\theta$ has to be considered in order to decide if the computed θ value goes above π or below ϕ, and care has to be taken to overcome zero denominators.

According to the notation of Fig. 1.14 the explicit finite difference scheme, centered in x, proposed for the solution of (1.61), (1.62) takes the form:

(1) Wave angle equation: solved for θ_{ij+1}

$$\theta_{ij+1} = \arccos\left[\left(\frac{\Delta y_j}{2}\left(\left(\frac{\sin\theta_{i+1j}}{c_{i+1j}} - \frac{\sin\theta_{ij}}{c_{ij}}\right)\frac{1}{\Delta x_i} + \left(\frac{\sin\theta_{ij}}{c_{ij}} - \frac{\sin\theta_{i-1j}}{c_{i-1j}}\right)\frac{1}{\Delta x_{i-1}}\right) + \frac{\cos\theta_{ij}}{c_{ij}}\right)\cdot c_{ij+1}\right] \quad (1.63)$$

(2) Wave height equation: solved for H_{ij+1}

$$H_{ij+1}^2 = \frac{1}{c_{gij+1}\sin\theta_{ij+1}}\left[(H^2c_g\sin\theta)_{ij} - \frac{\Delta y_j}{2}\left(\frac{(H^2c_g\cos\theta)_{i+1j}}{\Delta x_i} - \frac{(H^2c_g\cos\theta)_{ij}}{\Delta x_i} + \frac{(H^2c_g\cos\theta)_{ij} - (H^2c_g\cos\theta)_{i-1j}}{\Delta x_{i-1}}\right)\right] \quad (1.64)$$

The boundary conditions completing the model are:

(i) It is assumed that uniform conditions exist on the lateral boundaries (parallel depth contours).
(ii) The θ and H values are given as initial conditions on the open sea boundary ($j=1$).
(iii) The computation is terminated on the coastal boundaries ($h=0$).

The wave breaking criterion is applied in shallow waters. The computed wave height H_{ij} is compared to $0.78h_{ij}$; if $H_{ij} > 0.78h_{ij}$, then $H_{ij} = 0.78h_{ij}$.

The procedure is organised in BASIC in Program 4.

PROGRAM 4: WAVE REFRACTION MODEL. WAVE ANGLE AND HEIGHT COMPUTATION

```
>L.
    5REM WAVE REFFRACTION WAVE ANGLE AND HEIGHT CO
MPUTATION
   10DIMH(20,20),TH(20,20),C(20,20),N(20,20),L(20,
20),HH(20,20),K(20,20),S(20,20),LL(50)  ,DX(20),DY(
20)
   20READ PER,HHO,THO,IM,JM,HMAX
   30DATA...
   35FORJ=1TOJM:FORI=1TOIM:READH(I,J):NEXTI:NEXTJ

   40DATA...
   60FORI=OTOIM:READDX(I):NEXTI
   65DATA...
   70FORJ=OTOJM:READDY(J):NEXTJ
   75DATA...
   80LO=9.81*PER^2/6.28:CO=LO/PER
   85FORK=1TOHMAX:L=LO:TM=L
   90MN=(L+TM)/2:TM=L:A=2*PI*K/MN
   95L=LO*(EXP(A)-EXP(-A))/(EXP(A)+EXP(-A))
  100IFABS(L-TM)>.1 THEN GOTO90
  110LL(K)=L:LL(0)=0
  130NEXTK
  140FORI=1TOIM:FORJ=1TOJM
  150IFH(I,J)=0 THEN GOTO200
```

```
160FORK=0TOHMAX
170IFH(I,J)<K THEN GOTO190
180NEXTK
190L(I,J)=LL(K-1)+(LL(K)-LL(K-1))*(H(I,J)-K+1):K
(I,J)=2*PI/L(I,J)
192A=4*PI*H(I,J)/L(I,J):N(I,J)=.5*(1+A/(EXP(A)-E
XP(-A))*2)
195C(I,J)=CO*L(I,J)/LO
200NEXTJ:NEXTI
210FORI=1TOIM:FORJ=1TOJM:S(I,J)=K(I,J):NEXTJ:NEX
TI
220FORI=1TOIM:HH(I,1)=HH0:TH(I,1)=TH0:HH(I,2)=HH
0:TH(I,2)=TH0:NEXTI
310FORJ=3TOJM
320COR=0
340FORI=1TOIM
350IFH(I,J)=0 THEN GOTO430
360R=(S(I+1,J-1)*SIN(TH(I+1,J-1))-S(I,J-1)*SIN(T
H(I,J-1)))*DY(J-1)/S(I,J)/DX(I)
370L=(S(I,J-1)*SIN(TH(I,J-1))-S(I-1,J-1)*SIN(TH(
I-1,J-1)))*DY(J-1)/S(I,J)/DX(I-1)
380IF I=1 THEN L=R
390IF I=IM THEN R=L
400B=S(I,J-1)/S(I,J)*COS(TH(I,J-1))+(R+L)/2
410TH(I,J)=ACS(B)
420IF TH(I,J-1)>PI AND TH(I,J-1)<2*PI THEN TH(I,
J)=2*PI-TH(I,J)
430NEXTI
440FORI=1TOIM
450IF H(I,J)=0 THEN GOTO 510
455 IF SIN(TH(I,J))=0 THEN HH(I,J)=HH(I,J-1) :GO
TO510
460R=((HH(I+1,J-1)*C(I+1,J-1))^2*N(I+1,J-1)*S(I+
1,J-1)*COS(TH(I+1,J-1))-(HH(I,J-1)*C(I,J-1))^2*N(I
,J-1)*COS(TH(I,J-1))*S(I,J-1))*DY(J-1)/DX(I)
470L=((HH(I,J-1)*C(I,J-1))^2*N(I,J-1)*S(I,J-1)*C
OS(TH(I,J-1))-(HH(I-1,J-1)*C(I-1,J-1))^2*N(I-1,J-1
)*S(I-1,J-1)*COS(TH(I-1,J-1)))*DY(J-1)/DX(I-1)
480IF I=1 THEN L=R
490IF I=IM THEN R=L
500HH(I,J)=SQR(((HH(I,J-1)*C(I,J-1))^2*S(I,J-1)*
N(I,J-1)*SIN(TH(I,J-1))-(R+L)/2)/C(I,J)^2/S(I,J)/N
(I,J)/SIN(TH(I,J)))
505IFHH(I,J)>.78*H(I,J) THEN HH(I,J)=.78*H(I,J)
510NEXTI
520NEXTJ
530CLS:FORJ=1TOJM:FORI=1TOIM:MOVE(I-1)*80+100,(J
-1)*80+100:DRAW(I-1)*80+100+HH(I,J)*50*COS(TH(I,J)
),(J-1)*80+100+HH(I,J)*50*SIN(TH(I,J)):NEXTI:NEXTJ

540END
```

Description of main variables:

PER	= wave period
HH0	= wave height in the open sea ($J = 1$)
TH0	= wave angle in the open sea ($J = 1$)
IM, JM	= ranges of i, j indices along Ox, Oy
HMAX	= maximum water depth in the area in m (integer)
H(I, J)	= water depth at the grid nodes
DX(I), DY(J)	= variable discretisation steps along Ox, Oy respectively

The application refers to a coastal region having a hump in its bathymetry near the coastline. At a distance from the coast the isodepths become straight parallel lines. The wave orthogonals are normal to the isodepths in the open sea but are bent, focussing over the submarine hump. The bathymetry and wave conditions are symmetric. The flow domain and its discretisation are depicted in Fig. 1.15.

The application data are PER = 10 s, HH0 = 1 m, TH0 = 1.5708 rad, IM = JM = 11, HMAX = 11, DX = 10 m, DY = 10 m. The resulting H vectorised values are plotted in Fig. 1.15.

1.6.3 Wave breaking near the coast

Waves, considered as a hydrodynamic disturbance, become unstable in the open sea (no bottom influence) for a certain limiting value of the steepness H/L and in shallow waters for a certain limiting value of the relative height H/h. The linear hydrodynamic stability theory states that the maximum possible steepness is

$$\left(\frac{H}{L}\right)_{\max} = \tfrac{1}{7}\tanh\frac{2\pi h}{L} \tag{1.65a}$$

The stability consideration for solitary waves (approximating quite well to waves in shallow water) gives the maximum H/h value

$$\left(\frac{H}{h}\right)_{\max} = 0.78 \tag{1.65b}$$

or more generally

$$\frac{H_{br}}{h} = \gamma \quad (\gamma = \text{breaking index}) \tag{1.65c}$$

It has become evident from experiments that the breaker index is a function of the form of the breaking wave; it is a function of the steepness and the bed slope and varies from 0.5 to 1.2, with mean

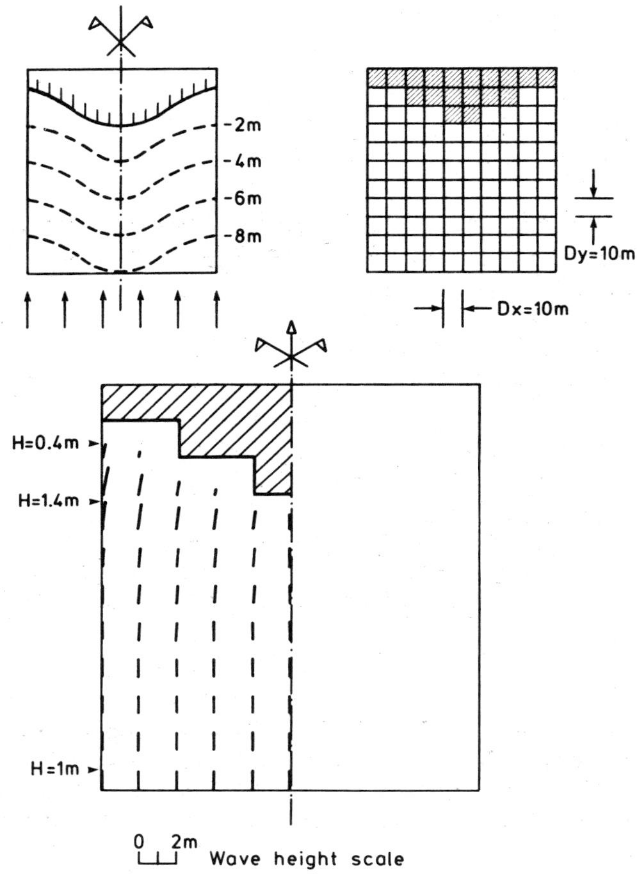

Fig. 1.15 Wave refraction. Flow domain, discretisation and computed H, θ (vectoral form)

value 0.7. In practice, the breaking depth and breaker height can be obtained from the wave steepness and the bed slope using Goda's diagrams (Fig. 1.16).

The open sea height H_0 refracted to H_0' is used in these diagrams. First, the breaking depth is assumed, the refraction is computed and the breaker height H_b is estimated. This is followed by estimation of a new breaking depth h_b'. If this is different from the assumed h_b the procedure is repeated with h_b' until convergence is satisfactory.

Example: To compute the breaker height and breaking depth on a beach of bed slope 1:20. Assuming that the wave height after the breaking point is decreasing linearly with depth, find the wave height

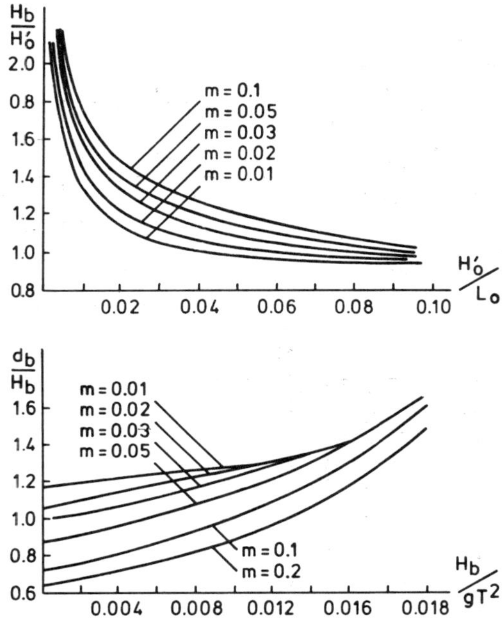

Fig. 1.16 Wave breaking diagrams

at a depth = 1.5 m. $H_0 = 2$ m, $T = 7$ m. The wave propagates normal to the coast ($H_0' = H_0$). First, $L_0 = gT^2/2\pi = 76.5$ m and $H_0/L_0 = 0.026$ are computed. For $m = 0.05$, from the first diagram of Fig. 1.16, it is found that $H_{br} = 1.3$. From the second diagram of Fig. 1.16,

$$H_b/gT^2 = 0.0054 \rightarrow h_{br} = H_{br} \cdot 1 = 2.6 \text{ m}$$

In this case $\gamma = 1$. At a depth $h = 1.5$ m, the wave height is $H_{1.5} = 1.5$ m.

1.6.4 Wave energy dissipation due to bed friction

For values of $HL/h^2 < 20 - 30$ the oscillating boundary layer near the bed developing under a progressive wave is a very small fraction of the water depth and the resulting frictional losses are negligible. For larger values, the loss of energy resulting in reduction of the wave height is considerable for long travel distances. Experimental evidence shows that the relation between the boundary layer thickness δ, the absolute bed roughness k_N and the amplitude of

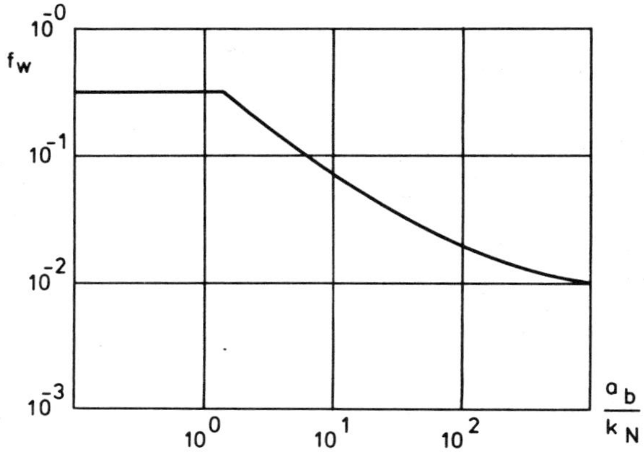

Fig. 1.17 Variation of bed friction coefficient f_w with roughness and wave induced orbital amplitude

oscillation of the water particles near the bed a_b is

$$\frac{30\delta}{k_N}\log\frac{30\delta}{k_N}=1.2\frac{a_b}{k_N} \tag{1.66a}$$

where

$$a_b \simeq \frac{HT}{4\pi}\sqrt{\left(\frac{g}{h}\right)} \quad \text{(for shallow water)} \tag{1.66b}$$

The above relations indicate the smallness of the δ/h ratio. For example, for $H=2$ m, $h=10$ m, $k_N=5$ cm and $T=60$ s, it is found that $\delta=11$ cm. The rate of energy loss due to bed friction is given by the classical relation of hydrodynamics

$$E_{df}=\frac{2}{3\pi}\rho f_w u_{b_{\max}}^3 \tag{1.67}$$

where E_{df} units are energy/unit bottom surface/second, $u_{b_{\max}}$ is the amplitude of the velocity near the bed, according to 1st order theory.

$$u_{b_{\max}}=\frac{H}{2}\sqrt{\left(\frac{g}{h}\right)} \quad \text{(for shallow water)} \tag{1.68}$$

and f_w the bed friction coefficient relating the bed friction to the velocity amplitude

$$\tau_b=\frac{f_w}{2}\rho u_{b_{\max}}^2 \tag{1.69a}$$

The friction coefficient according to the analysis by Jonsson is related to the a_b/k_N ratio

$$f_w = 0.3 \qquad \text{for } a_b/k_N < 1.57 \tag{1.69b}$$

$$\frac{1}{4\sqrt{(f_w)}} + \log\frac{1}{4\sqrt{(f_w)}} = -0.08 + \log\frac{a_b}{k_N} \qquad \text{for } \frac{a_b}{k_N} > 1.57 \tag{1.69c}$$

A graphical representation of the equations is given in Fig. 1.17.

1.7 RADIATION STRESS

In Chapter 2, where wave induced circulation models will be presented, the contribution of waves to mass transport in the coastal areas will be analysed in detail. At present, it is convenient to investigate how a progressive wave contributes, through the induced horizontal momentum and pressure components, to the dynamic equilibrium of a water column and to define and formulate the radiation stress magnitudes. These magnitudes will be used in the circulation models.

Independently of first order wave theory postulation that waves transport no mass in the direction of their propagation due to the periodicity and symmetry of the u magnitude, there is a surplus of momentum flux showing that a progressive wave can induce circulation through spatial gradients of induced mean momentum. The normal component of the radiation stress along the Ox direction of wave propagation is defined as:

$$\sigma_{xx} = \frac{1}{T}\int_0^T \mathrm{d}t\left[\rho\int_{-h}^{n} u^2\,\mathrm{d}z + \int_{-h}^{n} p\,\mathrm{d}z - \tfrac{1}{2}\rho g h^2\right]$$
$$= M + P - \tfrac{1}{2}\rho g h^2 \tag{1.70}$$

Substitution of the u function from first order wave theory (Equation (1.6)) ($O[u] \propto H$ and $O[u^2] \propto H^2$), gives for the M component of σ_{xx}

$$M = \frac{\gamma H^2}{16}(1+G) = \frac{\bar{E}}{2}(1+G) \tag{1.71}$$

where

$$G = \frac{2kh}{\sinh 2kh} = 2n - 1$$

For the estimation of the pressure component the p function deriving from the Stokes second order theory has to be used (Equation (1.15)) so that the σ_{xx} expression retains the same order of magnitude $O[H^2]$. Simple mathematical manipulation gives

$$P - \tfrac{1}{2}\rho g h^2 = \frac{\gamma H^2}{16} G = \frac{\bar{E}}{2} G \tag{1.72}$$

The sum of the two components (1.71) and (1.72) is the value of the main radiation stress component in the direction of the wave propagation

$$\sigma_{xx} = \frac{\gamma H^2}{16}(1 + 2G) = \frac{\bar{E}}{2}(1 + 2G) \tag{1.73}$$

Normal to that direction, there is only the pressure component, as the velocity v along Oy is zero in the assumed long crested waves situation.

$$\sigma_{yy} = \frac{\bar{E}}{2} G \tag{1.74}$$

The radiation stress magnitude behaves with respect to the coordinate transformation as a 2nd order tensor. Thus, in a coordinate system $x'Oy'$ making an angle θ with the principal axes xOy, Euler's law of transformation of coordinates gives

$$\sigma_{x'x'} = \sigma_{xx}\cos^2\theta + \sigma_{yy}\sin^2\theta - 2\sigma_{xy}\sin\theta\cos\theta \tag{1.75a}$$

$$\sigma_{y'y'} = \sigma_{xx}\sin^2\theta + \sigma_{yy}\cos^2\theta + 2\sigma_{xy}\sin\theta\cos\theta \tag{1.75b}$$

$$\sigma_{x'y'} = (\sigma_{xx} - \sigma_{yy})\sin\theta\cos\theta + \sigma_{xy}(\cos^2\theta - \sin^2\theta) \tag{1.75c}$$

Substitution of (1.73), (1.74) in (1.75a, b, c), given $\sigma_{xy} = 0$, leads to the following forms for the components of the radiation stresses with respect to a system of coordinates making an angle θ with the principal axes (wave propagation $\equiv$ positive Ox).

$$\sigma_{x'x'} = \frac{\bar{E}}{2}(2n - 1) + \bar{E}n\cos^2\theta \tag{1.76}$$

$$\sigma_{x'y'} = \sigma_{y'x'} = \frac{\bar{E}}{2} n \sin 2\theta \tag{1.77}$$

$$\sigma_{y'y'} = \frac{\bar{E}}{2}(2n - 1) + \bar{E}n\sin^2\theta \tag{1.78}$$

according to the notation of Fig. 1.18.

The induced set-up and set-down of the mean surface level due to waves normal to a coast will be analysed here as the most simple realisation of the wave contribution to coastal circulation. Using the notation of Fig. 1.19, if the temporal-mean depth-integrated

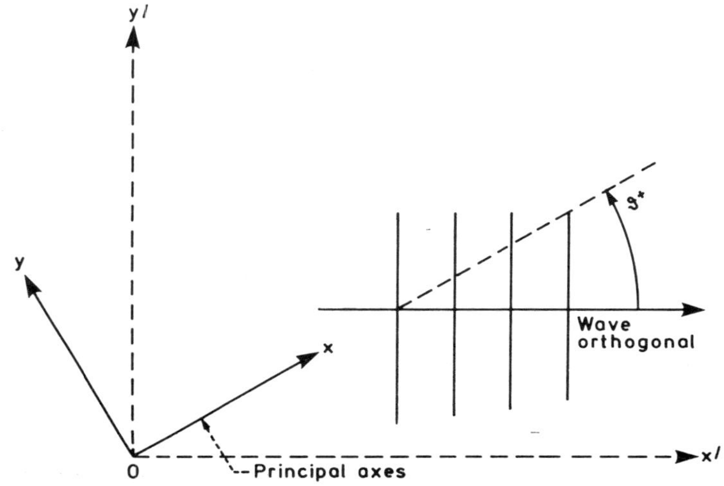

Fig. 1.18 Coordinate axes for radiation stresses formulation

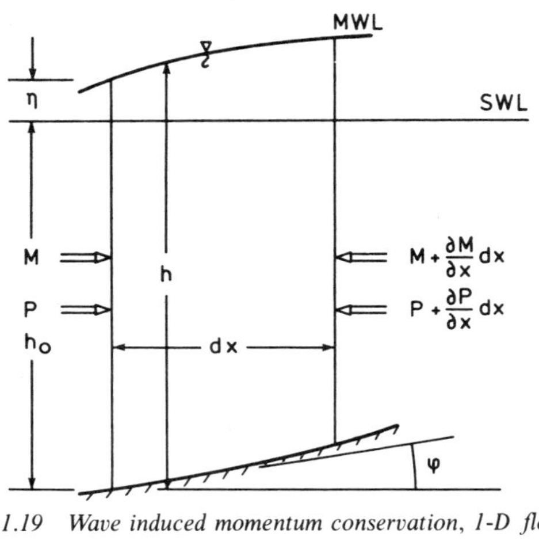

Fig. 1.19 Wave induced momentum conservation, 1-D flow

momentum and pressure components induced by the waves are denoted by M and P, the equilibrium of forces in the x direction leads to the equation:

$$\frac{\mathrm{d}}{\mathrm{d}x}(M + P) + gh\,\mathrm{tg}\,\varphi = 0 \tag{1.79}$$

Substituting (1.70) in (1.79) and since

$$h = h_0 + \eta, \qquad \mathrm{tg}\,\varphi = -\mathrm{d}h_0/\mathrm{d}x \tag{1.80}$$

it is finally found that:

$$\frac{d\sigma_{xx}}{dx} = -\gamma h \frac{d\eta}{dx} \tag{1.81}$$

Equation (1.81) shows that the mean water level varies normal to the coast inversely as the radiation stress σ_{xx}. The increase in σ_{xx} from the open sea to the breaking point results in a slight set-down of the sea level, while the decrease of σ_{xx} from the breaking line to the coast, in combination with the decrease of depth, results in a considerable set-up of the mean sea level.

In the following example, Equation (1.81) is solved by a simple finite difference scheme, on a domain discretised via a Δx step. On each cross section, σ_{xx} is computed from the depth by means of (1.73). Then (1.81) is used to compute η_{i+1} from the known h_i. The integration starts from a depth h_{01} where $\eta_1 = 0$ (deep water). The scheme can be explicit if $h \approx h_0$ or implicit if $h = h_0 + \eta$. A more general BASIC code (Program 5) giving the mean sea level evolution for arbitrary depth profile and given wave elements H, T, comprises the computation of the wave heights for various depths h_i, the subsequent computation of L_i, H_i, σ_{xx_i} and the final computation of η_i. Inside the breaker zone the wave height is estimated by selecting a breaker index γ.

PROGRAM 5: WAVE SET-DOWN, SET-UP, NORMAL TO THE COAST

```
10 REM WAVE SET DOWN SET UP NORMAL TO A COAST
20DIM L(50),D(50),N(50),H(50),E(50),SX(50),Z(50
)
30READ T,HO,IM
40DATA...
50FORI=1TOIM:READ D(I):NEXTI
60DATA...
70LO=9.81*T^2/2/PI
80FORI=1TOIM:MN=LO:IT=0:L(I)=LO
82IFD(I)>0 THEN GOTO 90
84L(I)=0:N(I)=1:H(I)=0:GOTO 170
90IT=IT+1:IF IT>100 THEN GOTO 210
100A=EXP(2*PI*D(I)/MN):B=EXP(-2*PI*D(I)/MN):TP=L
(I)
110L(I)=LO*(A-B)/(A+B):IFABS(L(I)-TP)<.001 THEN
GOTO 130
120MN=(TP+L(I))/2:GOTO90
130A=4*PI*D(I)/L(I):N(I)=.5*(1+A/(EXP(A)-EXP(-A)
)*2)
140H(I)=HO*SQR(.5*LO/L(I)/N(I))
150IF H(I)/D(I)<.8 THEN GOTO 170
```

```
  160H(I)=.8*D(I)
  170NEXTI
  180FORI=1TOIM:E(I)=9.81*H(I)^2/8:SX(I)=E(I)*(2*N
(I)-.5):NEXTI
  190Z(1)=0:FORI=2TOIM:Z(I)=Z(I-1)-2/9.81/(D(I)+D(
I-1))*(SX(I)-SX(I-1)):NEXTI
  200FORI=1TOIM:PRINTZ(I),L(I),N(I),H(I),E(I),SX(I
):NEXTI
  210END
```

Description of variables:

T = waves period in seconds
H0 = wave height for deep water in metres
IM = number of computation sections
D(I) = water depth corresponding to section #I
Z(I) = sea level elevation in section #I

The example refers to a coast of constant slope, 1:20, and waves characterised by $T = 6$ s and $H_0 = 2$ m. Irregularly spaced sections are examined (IM = 14). The water depth at the 14 cross-sections are:

$$D(I) = 25, 20, 15, 10, 9, 8, 7, 6, 5, 4, 3, 2, 1, 0 \text{ m}$$

The computed L, H, Z values are:

L (m)	55.8	55.0	53.1	48.4	46.9	45.2	43.2	40.9	38.1	34.8	30.7	25.8	—	—
H (m)	1.97	1.93	1.88	1.83	1.83	1.83	1.84	1.85	1.89	1.94	2.02	2.18	1.09	0
$Z \times 10^2$ (m)	0	−0.07	−0.25	−0.75	−0.95	−1.21	−1.56	−2.07	−2.85	−4.17	−6.58	−15.7	28.6	72.9

The flow domain discretisation and the graphical presentation of the mean sea level profile revealing the wave set-down and set-up are shown in Fig. 1.20.

1.8 WIND GENERATED WAVES

1.8.1 Wave hindcasting from wind data

The wave climate is an indispensable factor in the design of coastal structures. In most coastal locations in the world no wave records deriving from field measurements are available and both the time scale for the design and construction of technical works and the financial resources available make the installation and operation of such devices unfeasible for at least one climatological year. The most common, realistic practice is to base the wave prediction on wind measurements available from an appropriate coastal meteorological station. The meteorological station is selected to be characteristic of

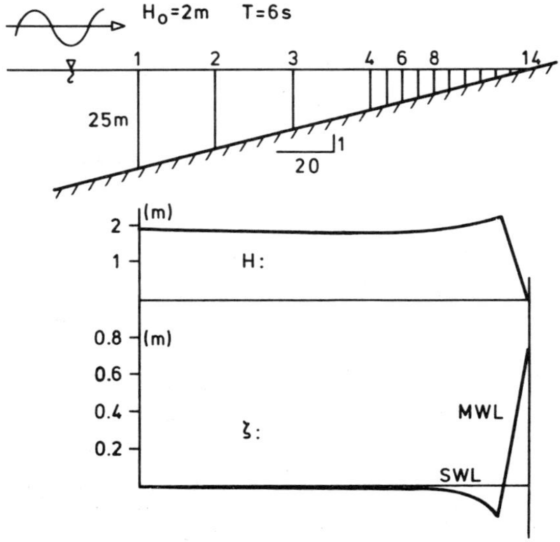

Fig. 1.20 Flow domain and wave induced set-down, set-up

the geographical domain under consideration, having first checked the wind records for their accuracy.

After selection of the station, one of the available methods for the prediction of the wave characteristics H and T from the wind velocity W, the wind duration t_D and the extent of the wave generation area (fetch) F can be applied. The wind data can be processed in tabular form, giving the annual frequency for various wind intensities and directions. Information is not usually available for the wind duration; for this one would need to have unprocessed data in the form of periodic observations of wind intensity and direction (every 3 or 6 hours) or continuous recordings. If the wind duration is missing, one must assume wave generation under given conditions of F and W and unlimited t_D, and then calculate the minimum necessary duration $t_{D_{min}}$ for H and T. The estimated value $t_{D_{min}}$ is checked to determine whether it is realisable, and if it is not, feedback on the H and T values may be necessary.

The available methods for the prediction of waves give the deep-water H_0 values. It is necessary, therefore, that the computed values undergo shoaling and refraction, to obtain transformed H values in shallow water.

The Darbyshire–Draper method of hindcasting waves from wind measurements is selectively presented here, as it is based on a sufficient number of field observations and as it distinguishes two cases, generation in oceanic waters and generation in coastal waters (~ 40 m depth). The method gives the value of H_{10}, the maximum wave height in a 10 min recording, related to the H_s (significant wave height, defined in Section 1.8.2), $H_{10} = 1.6H_s$ and the T_s (significant wave period) value close to the peak energy period. Figures 1.21 to 1.24 give the H_{10} and T_s values for deep and shallow waters.

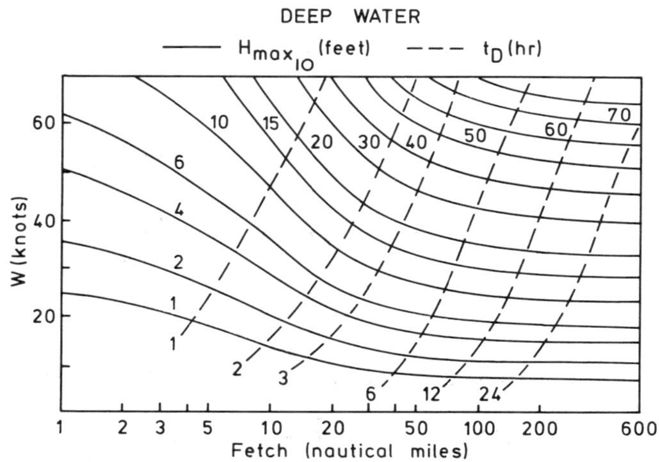

Fig. 1.21 Darbyshire-Draper diagram for H_s (deep water)

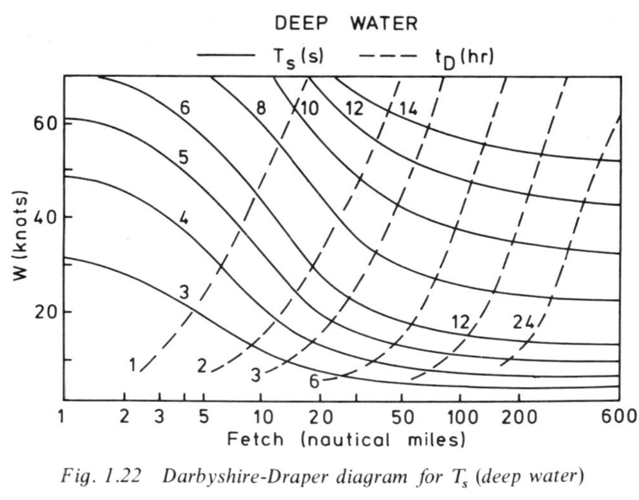

Fig. 1.22 Darbyshire-Draper diagram for T_s (deep water)

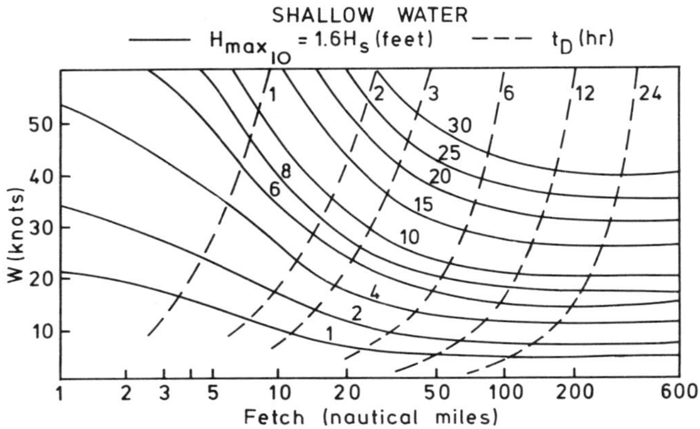

Fig. 1.23 Darbyshire-Draper diagram for H_s (shallow water)

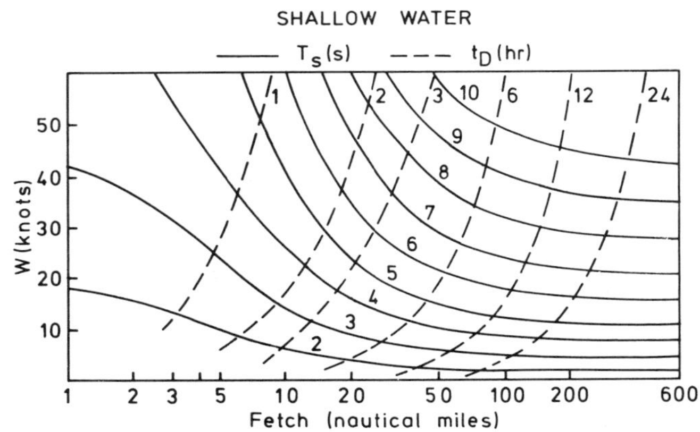

Fig. 1.24 Darbyshire-Draper diagram for T_s (shallow water)

1.8.2 Probabilistic analysis of wind generated waves

It is a common experience that in a real sea, the wind generated waves are not periodic and regular. The sea surface cannot be approximated by the simple trigonometric curve assumed by first order theory. The waves are rarely long-crested. The analysis of a recorded signal of sea surface elevations shows that many harmonic components are included in the waveforms with a variety of amplitudes and periods. The range of variations in the wave periods for wind generated wave episodes is not wide and a 'narrow band stochastic process' approximation is realistic.

The experimental analysis of the frequencies of appearance and the probabilities of exceedence of the relative maxima (amplitudes of waves) from a recorded signal, indicates that wave heights, considered as random variables, follow a Rayleigh distribution. This is a monoparametric probabilistic law, giving the probability of exceedence of a certain value H_i of wave height as

$$P(> H_i) = \mathrm{e}^{-(H_i/H_{rms})^2} \quad (1.82)$$

where H_{rms} is the root mean square value of the recorded wave heights. An estimate of the H_{rms} from a record of wave heights measured or predicted is given by

$$H_{rms} = \sqrt{\left(\frac{\sum_{i=1}^{N} H_i^2}{N-1}\right)} \quad (1.83)$$

The significant wave height H_s is defined as the mean value of the higher 33% of the wave heights. This magnitude is related to H_{rms} by the relation $H_{rms} = H_s/\sqrt{2}$ (with the result that (1.82) can be written as $P(> H_i) = \mathrm{e}^{-2(H_i/H_s)^2}$). It has an important physical meaning as it has been found to be the wave height that an experienced observer attributes to a confused wavy sea, by simple observation.

The maximum probable wave height estimated from a record containing N waves and characterised by significant wave height H_s is found to be, by use of Equation (1.82),

$$H_{\max} = \frac{1}{\sqrt{2}} H_s \sqrt{(\ln N)} \quad (1.84)$$

The Rayleigh distribution describes also the T^2 random variable

$$P(> T_i^2) = \mathrm{e}^{-2(T_i/T_s)^2} \quad (1.85)$$

From (1.85) it becomes clear that the range of T_i values is narrower than the range of H_i (narrow-band process assumption).

The probabilistic analysis presented refers to a wave episode lasting some hours and generated under specific meteorologic conditions. For example, if a wave recording lasts 20 min and the mean period is found to be $T = 5$ s, the Rayleigh distribution, etc. holds for a specific $20 \cdot 60/5 = 240$ pulses. Such recordings, either periodic or not, are recorded at coastal stations for one or more years. The distribution holding for the sample of H_s or $H_{\max}$ values corresponding to each recording, is found to be a specific form of Weibull distribution

$$P(> H_s) = \mathrm{e}^{-[(H_s + c)/a]^b} \quad (1.86)$$

with special values, $b = 1$, $c = 0$, i.e.

$$\ln P(> H_s) = -\frac{H_s}{a} \tag{1.87}$$

The undetermined parameter is easily found from a sequence of available H_s values (computed from a sequence of waves predicted or recorded) by a graphical representation on a semilogarithmic paper, with linear scale for the wave heights and logarithmic scale for the experimental frequencies of exceedence.

A very common problem arising in the design of coastal structures can be solved by means of Equations (1.82) and (1.87). Some specific, inflexible or important (from the safety point of view) parts of maritime structures have to be designed to sustain the hydrodynamic conditions entailed by a wave having a certain return period. The stability of a breakwater with vertical walls or of an element of an offshore structure has to be designed under the wave height with return period of 100 or 500 years. The following procedure is realised if only wind processed data are available (the worst case):

(1) The significant wave heights are computed from the various wind intensities and fetches. These heights are characterised by a known frequency of reappearance within the year, and consequently by a known probability of exceedence (cumulative frequency).

(2) A straight line is drawn relating the H_{s_i} values to their $\ln P(> H_{s_i})$ on a semilogarithmic paper.

(3) The H_s value with return period of 100 years, for example, is estimated from the corresponding $\ln P$. To achieve this, the duration of the episode $\# i$ characterised by H_{s_i} must be known. If the wind observations are periodic (every 3 hours or every 6 hours) it is assumed that the corresponding wind intensity characterises the interval between two measurements. Let us suppose that the measurements are done every 6 hours. There are $365 \times 24/6 = 1460$ such intervals within one year. So the wave height value with return period of 100 years has a probability of exceedence, $1/146{,}000 = 6.8 \times 10^{-6}$. The $H_{s_{100}}$ is estimated for that P value.

A distinction has to be made between the return period T of a wave of given height and the probability of its exceedence during a lifetime N years of a maritine structure. A wave with a period of recurrence T years is expected to occur with a mean frequency $1/T$ (or A/T times in a record of duration A years). The probability of exceedence of that wave once during N consecutive years is

$$P = 1 - (1 - 1/T)^N$$

so a wave of return period 1000 years has a probability of exceedence in a lifetime of 100 years equal to

$$P = 1 - (1 - 1/1000)^{100} = 9.5\%$$

(4) Based on that H_s value, the H_{max} (maximum probable value during the episode characterised by $H_{s_{100}}$) can be computed via Equation (1.84). The number N of waves in that episode can be computed from the assumed duration (6 hours) and a period characterising the episode. That can be the maximum hindcast T_s or the period found by use of an empirical formula, such as

$$\bar{T} = 4 + 2 \cdot H_s^{0.7} \tag{1.88}$$

holding in the Mediterranean Sea. Its exact value has not much importance as the number $N(= 3600 \times 6/T)$ enters in (1.84) as $\sqrt{(\ln N)}$.

An example of wave hindcasting and probabilistic estimation of extreme value done manually and organised in BASIC code will illustrate the above analysis.

Assume that an element of a coastal structure has to be designed for a 10 year period wave. The $H_{max_{10}}$ value has to be estimated on given wind data assuming no refraction and shoaling. The wind intensity-direction-frequency data are (Table 1.3):

Table 1.3

	Direction (fetch in km)			
Intensity	*North (40)*	*Northeast (30)*	*Northwest (60)*	*W (m/s)*
Weak	5%	12%	8%	7
Moderate	10%	3%	2%	15
Strong	1%	0.5%	0.5%	22

The H_s and T_s values are found (Table 1.4) from Darbyshire diagrams for coastal waters and fetches, 21.5 mile, 16 mile and 32.5 mile for the N, NE, NW winds, respectively:

Table 1.4

	North	*Northeast*	*Northwest*
Weak	0.6 m/3.5 s	0.4 m/3.5 s	0.7 m/4.0 s
Moderate	1.7 m/5.5 s	1.4 m/5.0 s	2.1 m/6.0 s
Strong	3.0 m/7.0 s	2.4 m/6.0 s	3.8 m/8.0 s

The corresponding $t_{D_{\min}}$ (minimum wind durations) for the generation of such waves are 4.5, 2.5, 2.0, 3.0, 2.0, 1.7, 4.5, 3.5 and 2.5 hours. These are considered realisable and no reduction of values is made, on wind duration limitation grounds. The listing of H_s values in ascending order, and the respective frequency and probability of exceedence, are given in Table 1.5.

Table 1.5

H_{s_i}	f_i	$P(\geqslant H_{s_i})$	$\ln P(\geqslant H_{s_i})$
0.4	0.12	0.42	−0.867
0.6	0.05	0.30	−1.204
0.7	0.08	0.25	−1.386
1.4	0.03	0.17	−1.772
1.7	0.10	0.14	−1.996
2.1	0.02	0.04	−3.218
2.4	0.005	0.02	−3.912
3.0	0.01	0.015	−4.199
3.8	0.005	0.005	−5.298

$\sum f_i = 0.42$

For a probability of exceedence $P(>H_{s_{10}}) = 68 \times 10^{-6}$ and $\ln P(>H_{s_{10}}) = -9.59$, it is found by extrapolation of the straight line drawn in Fig. 1.25 that $H_{s_{10}} = 7.3$ m. Application of Equation (1.88) gives $T = (4+2)H_s^{0.7} = 12$ s. Application of Equation (1.84) for

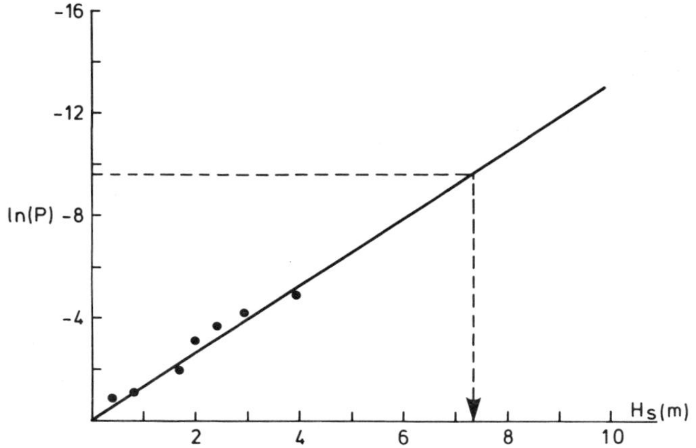

Fig. 1.25 Weibull distribution for the estimation of wave height H_s with period 10 ye

$N = 3600 \times 6/12 = 1800$ waves, gives

$$H_{\max_{10}} = \frac{1}{\sqrt{2}} \times 7.3\sqrt{\{\ln(1800)\}} = 14.2\ \mathrm{m}$$

This is quite a high value and it must be checked to determine if it is realisable in that coastal area or if wave breaking reduces it. For example $T = 12$ s corresponds to $L = 156$ m for a depth $h = 30$ m. Applying in an approximate way the breaking criterion, $(H/L)_{\max} = \frac{1}{7}$, it is found that $H_{\max} = 25$ m, and for $(H/h)_{\max} = 0.7$, that $H_{\max} = 21$ m, so the computed value $H_{\max_{10}}$ may be realised.

The procedure followed after hindcasting H_s, T, i.e. ordering of H_s, computation of experimental frequency of exceedence, computation of the slope of the Weibull line by the least squares method, computation by extrapolation of $H_{s_{10}}$ values and computation of the $H_{\max_{10}}$ value for the computed T and assumed 6 hours duration, is organised in BASIC and presented in Program 6.

PROGRAM 6: EXTREME WAVE HEIGHT PREDICTION. WEIBULL DISTRIBUTION FOR H_s

```
   10REM EXTREME WAVE HEIGHT PREDICTION WEIBULL DI
STRIBUTION FOR HS
   20DIMHS(20),T(20),F(20),FF(20)
   30READ W,PER
   35DATA...
   40FORI=1TOW:READHS(I),T(I),F(I):NEXTI
   50DATA...
   60TMAX=0:FORK=1TOW:IFTMAX>T(K) THEN TMAX=T(K) :
NEXTK
   70FORN=1TOW:HMAX=0
   80FORK=N TO W
   90IF HMAX>HS(K) THEN GOTO 110
  100HMAX=HS(K):KK=K:TEMP=F(K)
  110NEXTK
  120HS(KK)=HS(N):F(KK)=F(N):HS(N)=HMAX:F(N)=TEMP
  130NEXTN
  140SUM=0:FORN=1TOW:SUM=SUM+F(N):NEXTN
  150FF(W)=SUM:FORN=W-1 TO 1 STEP-1:FF(N)=FF(N+1)-
F(N+1):NEXTN
  160FORI=1TOW:PRINTHS(I),F(I),FF(I):NEXTI
  170FORN=1TO W:F(N)=LN(FF(N)):NEXTN
  180SUM1=0:SUM2=0:FORI=1TOW:SUM1=SUM1+HS(I)*F(I):
SUM2=SUM2+F(I)^2:NEXTI
  190A=SUM1/SUM2:FF=LN(1/PER/1460):HSS=A*FF
  200TM=4+2*HSS^.7:IF TM>TMAX THEN TMAX=TM
  210HMAX=.707*HSS*SQR(LN(3600*6/TMAX))
  220PRINTHMAX,TMAX
  230END
```

Description of main variables:

W = number of wave episodes with corresponding H_s, T and annual frequencies, available
PEK = period of reappearance of the required wave
Hs(I) = significant wave heights available
T(I) = corresponding periods
F(I) = corresponding annual frequencies

The application is performed with the previously found data, HS, T, F (hindcast wave heights and periods and annual frequencies respectively). It gives as slope of the straight line found by least squares, -0.6878. It is found that for PER = 10 years, $H_{s_{10}} = 6.6$ m, $H_{max_{10}} = 12.9$ m and $T = 11.5$ s.

2

Mathematical models of coastal circulation

2.1 DEFINITION—THE GENERAL FORM OF THE MODEL

Coastal circulation is defined by the development of generally non-steady velocity and surface elevation fields in a coastal geophysical domain where the depths are of the order of 10 or more metres, the horizontal dimensions are of the order of 10 or more kilometres and the geometry of the coastline is not simple. This geophysical domain is connected to the open sea through one or more openings (open sea boundaries). The circulation in these areas is generated and sustained by various generating factors such as the tide, the wind or atmospheric pressure acting on the water surface, the horizontal variation of wave momentum due to diffraction refraction and shoaling, and by the spatial variation of water density.

The defined coastal domain extends from the coastline to the continental slope, so it comprises of marine areas where currently engineering developments are extensive.

The definition given for the circulation applies to the waves described in Chapter 1. Indeed, the long-wave mathematical model will reappear in the present chapter, but the phenomena to be investigated here are differentiated from those of the first chapter, so far as their time scale is concerned (hours or days compared to seconds or minutes in the first chapter).

In the following sections, the phenomena and the corresponding mathematical models will be presented according to the generating factors. This distinction has mathematical rather than physical meaning, as the various circulation generating factors coexist and are mingled in varying proportions. The general mathematical model will be formulated in the present section and specific forms of the general model in subsequent ones, derived under special simplifying assumptions.

The general model for coastal circulation is based on the following physical assumptions:

(1) As the horizontal dimension of the flow domain L is several orders of magnitude larger than the vertical dimension (depth H), the assumption of nearly horizontal flow is realistic. The horizontal velocity components u, v, are several orders of magnitude larger than the vertical component, w. This observation is generally valid except for certain minor regions of the flow domain; such as areas of sharp bed slopes ($> 1{:}5$) or areas where upwelling or fronts occur. The assumption of nearly horizontal flow contributes to a considerable simplification of the model as it excludes the vertical velocity component w from the main unknown functions and leads to a hydrostatic pressure distribution. Mathematically speaking, this assumption simplifies the form of the vertical momentum conservation equation to

$$w \approx 0 \rightarrow \frac{\partial p}{\partial z} = -\rho g \tag{2.1}$$

(2) The horizontal dimension of the flow domain is usually very large in comparison with the magnitude of the horizontal velocities developed within them and the time taken for circulation to develop may reach the order of some days. During that time, the effect on the flow domain of the earth's rotation is such that the contribution of the Coriolis force (at least of its horizontal components) cannot be neglected. The Coriolis mass force is expressed by the term $2\rho\mathbf{\Omega} \times \mathbf{V}$ where $\mathbf{\Omega}$ is the angular rotation of the earth vector and $\mathbf{V}$ the fluid velocity vector. The relation between the Coriolis and inertial forces can be expressed via the dimensionless number $T\Omega$ or $\Omega L/U$, where T is the time scale of evolution of the phenomenon, L the horizontal dimension of the flow domain, and U a characteristic velocity magnitude. The second number, known as the Rossby number, is an indicator of the importance of the Coriolis effect.

The horizontal components of the Coriolis force are given by the equations

$$f_{xc} = 2\Omega(\sin\varphi)v = fv \tag{2.2}$$

$$f_{yc} = -2\Omega(\sin\varphi)u = -fu \tag{2.3}$$

where f is the Coriolis coefficient and φ the geographic latitude of the domain. If the flow domain is of limited horizontal dimension, the mean φ value is used and the f coefficient is constant over the whole domain. Extension of the domain over several degrees of latitude shows the the β-effect (variation of f) has to be considered. Although the Coriolis effect is included in coastal circulation models the

curvature of the earth is not considered in most cases and the flow domain is approximated by its mercatorial projection (f plane).
(3) A flow domain of large dimension results in quite large Reynolds numbers, $O[\mathrm{Re}] > 10^4$, even for minimal velocity values (order of 1 cm/s). The flows are always turbulent. For turbulent stresses, the Boussinesq approximation is made, approximating the Reynolds stresses by the turbulent mean velocity gradient. The eddy viscosity coefficient appearing in this approximation develops generally in an anisotropic way, depending on the nature of the turbulence. For the simplest possible realistic turbulence closure in applications of physical oceanographic scale, two final assumptions are made:

(i) The eddy viscosity coefficients are differentiated between the horizontal v_h and vertical v dimensions.
(ii) Constant or variable values of v_h and v_v (in general $v_h \neq v_v$) are adopted. In the second case the functional forms of v_h, $v_v(x, y, z)$ are the simplest possible. Prandtl mixing length theory is commonly used or in special cases, requiring the most detailed vertical current profile description, k or $k - \varepsilon$ models for turbulence closure are used. The mathematical expressions for v_v in ascending order of complexity are:

$$v_v = \text{constant}$$

$$v_v = \lambda u h \frac{z}{h}\left(1 + \frac{z}{h}\right)$$

$$v_v = l_m^2 \left|\frac{\partial u}{\partial z}\right|$$

$$v_v = L_0 \cdot k^{1/2} \qquad \text{where } k = \tfrac{1}{2}\overline{(u'^2 + v'^2 + w'^2)} \tag{2.4}$$

and l_m, L_0 are the mixing and dissipation lengths respectively, functions of z. Figure 2.1 shows the most common morphologies for the v_v distribution.

Under the abovementioned assumptions, the coastal circulation model in its most general form is composed of the equations of equilibrium of forces in a horizontal dimension, and the mass continuity equation. According to the notation of Fig. 2.2 these have the form:

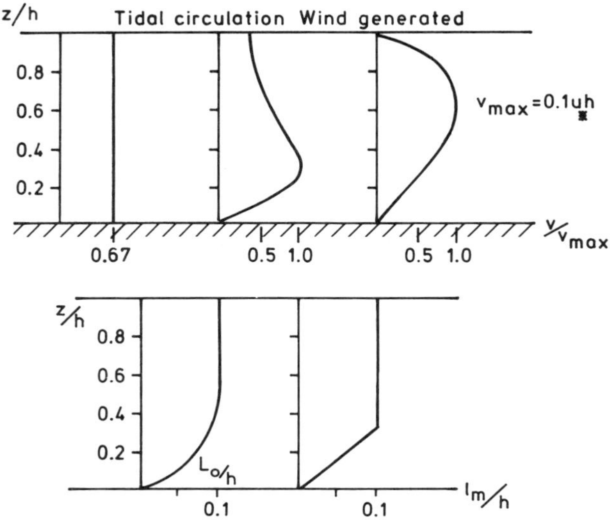

Fig. 2.1 Eddy viscosity and mixing length distributions for tidal and wind generated flows

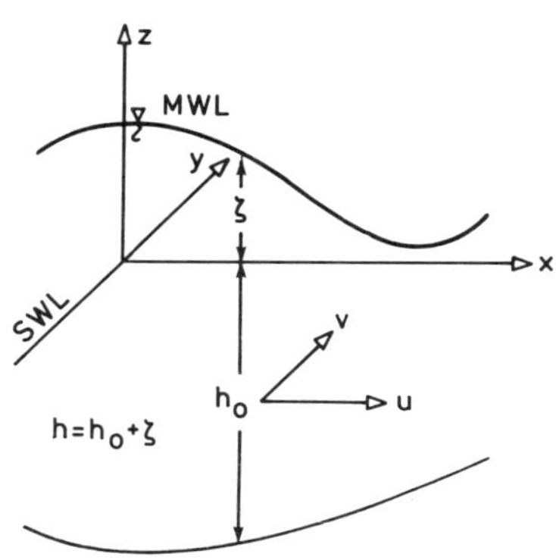

Fig. 2.2 Coordinate axes and basic symbols in circulation models

(1) Equilibrium equations:

$$\frac{\partial u}{\partial t}+u\frac{\partial u}{\partial x}+v\frac{\partial u}{\partial y}+w\frac{\partial u}{\partial z}$$
$$=-\frac{1}{\rho}\frac{\partial p}{\partial x}+a_x+\frac{\partial}{\partial x}\left(v_h\frac{\partial u}{\partial x}\right)+\frac{\partial}{\partial y}\left(v_h\frac{\partial u}{\partial y}\right)+\frac{\partial}{\partial z}\left(v_v\frac{\partial u}{\partial z}\right) \tag{2.5}$$

$$\frac{\partial v}{\partial t}+u\frac{\partial v}{\partial x}+v\frac{\partial v}{\partial y}+w\frac{\partial v}{\partial z}$$
$$=-\frac{1}{\rho}\frac{\partial p}{\partial y}+a_y+\frac{\partial}{\partial x}\left(v_h\frac{\partial v}{\partial x}\right)+\frac{\partial}{\partial y}\left(v_h\frac{\partial v}{\partial y}\right)+\frac{\partial}{\partial z}\left(v_v\frac{\partial v}{\partial z}\right) \tag{2.6}$$

where $u(x, y, z, t)$, $v(x, y, z, t)$, $w(x, y, z, t)$ and $p(x, y, z, t)$ are the velocity components and pressure functions and v_h, v_v the eddy viscosity functions.

(2) Equation of mass continuity (incompressible fluid):

$$\frac{\partial u}{\partial x}+\frac{\partial v}{\partial y}+\frac{\partial w}{\partial z}=0 \tag{2.7}$$

A more useful form for free surface, nearly horizontal flows, derives from the integration of (2.7) over the depth

$$\frac{\partial}{\partial x}\int_{-h_0}^{\zeta}u\,\mathrm{d}z+\frac{\partial}{\partial y}\int_{-h_0}^{\zeta}v\,\mathrm{d}z+w\Big|_{\zeta}-w\Big|_{-h_0}=0 \tag{2.8}$$

where $\zeta(x, y, t)$ is the free surface elevation relative to the still water level (SWL). For $w(x, y, z=-h)=0$ and $w(x, y, z=\zeta)\approx\partial\zeta/\partial t$ Equation (2.8) takes the useful form

$$\frac{\partial\zeta}{\partial t}+\frac{\partial}{\partial x}\int u\,\mathrm{d}z+\frac{\partial}{\partial y}\int v\,\mathrm{d}z=\frac{\partial\zeta}{\partial t}+\frac{\partial Uh}{\partial x}+\frac{\partial Vh}{\partial y}=q \tag{2.9}$$

where $U(x, y, t)$ and $V(x, y, t)$ are the depth mean horizontal velocity components and $q(x, y, t)$ ($[q]=L^3/L^2/T$) is the specific discharge of a source or sink that may exist in the flow domain, such as the effluence of a river, etc.

The variables a_x, a_y, describe distributed mass forces such as in our case the Coriolis horizontal components, or forces due to waves (gradients of radiation stresses that will be described in detail in Section 2.4 on wave generated circulation).

The vertical velocity component w appearing in the equilibrium equations may be either neglected or computed from the solution of the continuity equation (2.7).

The turbulence closure in coastal general circulation models is resolved via the approximation of v_v by one of the forms (2.4). A comparison of v_v and v_h and consequently of the horizontal and vertical momentum diffusion terms can be analysed as follows: As

$$[v]=L^2/T$$

then

$$O[v_h] \propto O[L] \cdot O[U]$$
$$O[v_v] \propto O[H] \cdot O[U]$$

giving

$$O[v_v] \ll O[v_h] \qquad \text{since} \qquad O[L] \gg O[H]$$

where L and H are the horizontal and vertical flow field dimensions. Expressions for the horizontal and vertical momentum diffusion are given by

$$O\left[\frac{\partial}{\partial x}\left(v_h \frac{\partial u}{\partial x}\right)\right] \propto \frac{O[L] \cdot O[U]^2}{O[L]^2} \propto \frac{O[U]^2}{O[L]}$$

and

$$O\left[\frac{\partial}{\partial z}\left(v_v \frac{\partial u}{\partial z}\right)\right] \propto \frac{O[H] \cdot O[U]^2}{O[H]^2} \propto \frac{O[U]^2}{O[H]}$$

The second expression dominates the first. The horizontal momentum diffusion terms thus may be neglected before the vertical terms. Their retention and use of constant v_h values is recommended for numerical stability reasons as the induced physical/numerical diffusion smooths out numerically produced perturbations in the velocity field.

The field equations described are completed by appropriate boundary conditions. From the physical point of view these can be distinguished as:

(1) The coastal perimeter. The velocity component normal to this boundary is suppressed. The tangential component is left free to develop, simulating any longshore current. The suppression of the normal velocity does not permit flooding of low coastal areas. In the case of strong storm surges, special weir type conditions have to be applied to the boundary to describe lowland flooding.

(2) The open sea. This boundary in itself has no strict character, no easily discernible features, but it is indispensable for reasons of computational economy. It is a physical or fictitious line separating the investigated domain from the rest of the sea. It is easily fixed in the case of a semi-enclosed coastal basin connected via a fixed opening to a wide, deep water body. If on that boundary no in situ measurements of velocity or sea surface elevation are given some assumptions have to be made. A most convenient condition is one describing the free radiation to the open sea (without back reflection) of any perturbation reaching that boundary from inside. The mathematical expression quantifying this physical operation derives from the conservation principle for a moving surge and relates the depth mean velocity U_n, normal to the boundary, to the free surface elevation ζ

$$\mathbf{n} \cdot \mathbf{U}_n = \zeta \sqrt{(g/h)} \tag{2.10}$$

where **n** is the unit outward vector.

Equivalent to Equation (2.10), but deriving from the theory of linear long waves moving unidirectionally, the following relation describes, in differential form, the evolution of the radiated part of the ζ magnitude

$$\frac{\partial \zeta_r}{\partial t} + \frac{\partial \zeta_r}{\partial n} \cdot \sqrt{(gh)} = 0 \tag{2.11}$$

The total value of ζ is assumed as the superposition of ζ_i(incident) + ζ_r(radiated)

$$\zeta = \zeta_i + \zeta_r \tag{2.12}$$

(3) The free surface. The wind, having near surface velocity components W_x, W_y (measured at a height of 10 m above the surface) exercises a frictional force on the surface of the water. The x, y components of that force can be expressed by the quadratic forms:

$$\frac{\tau_{sx}}{\rho} = \nu_v \left.\frac{\partial u}{\partial z}\right|_{z=\zeta} = kW_x\sqrt{(W_x^2 + W_y^2)} \tag{2.13}$$

$$\frac{\tau_{sy}}{\rho} = \nu_v \left.\frac{\partial v}{\partial z}\right|_{z=\zeta} = kW_y\sqrt{(W_x^2 + W_y^2)} \tag{2.14}$$

where k is a friction coefficient (with values $O[k] = 10^{-6}$) and ρ is the water density.

(4) The sea bed. On the sea bed or at a height $\sim k_s/30$ above, where k_s is the absolute bed roughness, the velocity may be assumed equal to zero. The 'no slip' condition is the simplest one. The shear stress developed near the bed is defined by the relations

$$\frac{\tau_{bx}}{\rho} = \nu_v \left.\frac{\partial u}{\partial z}\right|_{z=-h}, \qquad \frac{\tau_{by}}{\rho} = \nu_v \left.\frac{\partial v}{\partial z}\right|_{z=-h} \tag{2.15}$$

Instead of the 'no slip' condition the bed shear may be related to the velocity magnitude at a height Δz above the bed. This relation is based on the assumption that over the distance Δz the velocity follows the logarithmic distribution for uniform turbulent flow,

$$u = \frac{u_*}{k} \ln \frac{z+h}{z_0} \tag{2.16}$$

where $k = 0.4$, $u_* = \sqrt{(\tau_{bx}/\rho)}$, $z_0 = k_s/30$. At a distance Δz

$$\left. u \right|_{z=-h_0+\Delta z} = \frac{1}{k}\sqrt{\left(\frac{\tau_{bx}}{\rho}\right)} \ln \frac{\Delta z}{z_0} \tag{2.17}$$

Readers will recall that the velocity distribution for uniform flow (2.16) results in the following relation of τ_b/ρ to the depth mean velocity U:

$$\frac{\tau_b}{\rho} = U^2\left(\frac{k}{\ln(h/z_0) - 1}\right)^2 = \lambda U^2 = \frac{g}{C^2} U^2 \tag{2.18}$$

where λ is a nondimensional friction coefficient and C the Chézy bed friction coefficient.

2.2 MATHEMATICAL MODELS OF LONG WAVE INDUCED CIRCULATION (TIDAL MODELS)

2.2.1 Formulation in terms of the depth mean velocities

The circulation generating factor is a periodic or non-periodic perturbation of the free surface elevation, arriving from the open sea and developing over a period of several hours (in the most common case of the M_2 tide component generated by the influence of the moon on the earth, it is periodic with period 12.8 h). The long waves arriving from the open sea enter through the open sea boundary, propagate to the coastal area and are reflected from the coastal boundaries. They are subject to various deformations due to diffraction, refraction shoaling and frictional losses of energy and part of their energy is radiated through the open sea boundary back to the sea.

The applicability of the long-wave mathematical theory can be verified by comparison with the wave length L and the depth h. For $O[h] = 100$ m $(L \gg h)$. For wave amplitudes $O[H] = 1$ m, the ratio HL/h^2 is much greater than 20–30. The flow develops as a boundary layer $(h = \delta)$. As this phenomenon evolves slowly with time the velocity profile develops uniformly in stages,

$$u = \frac{u_*}{k} \ln \frac{h + z}{z_0} \tag{2.16}$$

The wave celerity is $O[C] = 30$ m/s, resulting, in the case of an M_2 tide, in a wave length $O[L] = 1.3 \times 10^6$ m.

The current intensity is almost uniform over the depth, steep gradients developing only near the bed. The uniformity of velocity over the depth permits simplification of the general model (2.5)–(2.7) by integration over the depth and the introduction of the depth mean velocity values:

$$U = \frac{1}{h}\int_{-h_0}^{\zeta} u \, dz, \qquad V = \frac{1}{h}\int_{h_0}^{\zeta} v \, dz \tag{2.19}$$

In the case of homogeneous fluid, the pressure terms become under the hydrostatic pressure approximation

$$-\frac{1}{\rho}\cdot\frac{\partial p}{\partial x}=-g\frac{\partial \zeta}{\partial x},\qquad \frac{1}{\rho}\cdot\frac{\partial p}{\partial y}=-g\frac{\partial \zeta}{\partial y} \tag{2.20}$$

The Coriolis terms are retained. The integration of (2.5), (2.6) over the depth, the approximation of the non-linear convective terms by their depth mean values and the expression of the bed friction using the quadratic forms (2.18) result in a simple model known as the 2DH (2-dimensional horizontal flow) model

$$\frac{\partial U}{\partial t}+U\frac{\partial U}{\partial x}+V\frac{\partial U}{\partial y}=-g\frac{\partial \zeta}{\partial x}+fV-\frac{gU\sqrt{(U^2+V^2)}}{hC^2}+\nu_h\nabla_h^2 U \tag{2.21}$$

$$\frac{\partial V}{\partial t}+U\frac{\partial V}{\partial x}+V\frac{\partial V}{\partial y}=-g\frac{\partial \zeta}{\partial y}-fU-\frac{gV\sqrt{(U^2+V^2)}}{hC^2}+\nu_h\nabla_h^2 V \tag{2.22}$$

$$\frac{\partial \zeta}{\partial t}+\frac{\partial Uh}{\partial x}+\frac{\partial Vh}{\partial y}=0 \tag{2.23}$$

where ∇_h^2 is the horizontal Laplacian, $\nabla_h^2=\partial^2/\partial x^2+\partial^2/\partial y^2$.

2.2.2 Numerical solution of the 2-D horizontal flow model by explicit finite differences

Various numerical schemes for the solution of the long wave induced circulation models in coastal domains are still the subject of investigation by researchers. The models actually treated here consist of Equations (2.21)–(2.23) in the given or slightly modified forms. The schemes are based on the finite difference or finite element methods. A finite difference scheme selected on the basis of its simplicity and efficiency is presented here.

The flow domain is discretised by an orthogonal horizontal grid with mesh sides $\Delta x, \Delta y$. The lateral coastal boundaries are approximated by mesh sides $\parallel Ox$, or $\parallel Oy$. The open sea boundaries are also selected piecewise by $\parallel Ox$, or $\parallel Oy$. The unknown functions U, V, ζ are computed on characteristic locations in a staggered way. The U, V values refer to the mesh sides $\parallel Oy$ and $\parallel Ox$ respectively and the ζ refer to the mesh centers. The coordinates of the computation points are characterised by the i, j, n indices, the first referring to the abscissa, the second to the ordinate and the third to the time. The

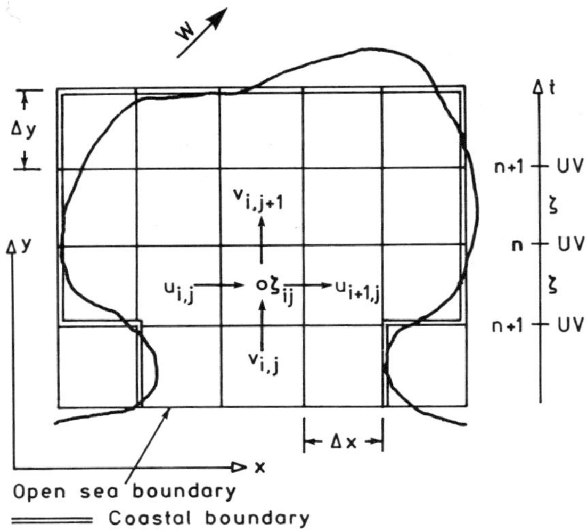

Fig. 2.3 Orthogonal staggered grid for spatial and time discretisations

computation points and the indices i, j relevant to a typical mesh are illustrated in Fig. 2.3.

The terms of the Equations (2.21)–(2.23) are approximated by finite differences. Forward differences are used for the time derivative and centered differences for the rest of the space derivatives, synthesising an explicit numerical scheme. The computation of U, V values at the $n+1$ time level involves known values of U, V, ζ (n and $n+\frac{1}{2}$ level) and there is no algebraic system to be solved.

The functional forms

$$U_{ij}^{n+1} = f_1(U^n, V^n, \zeta^{n+\frac{1}{2}}) \tag{2.24}$$

$$V_{ij}^{n+1} = f_2(U^n, V^n, \zeta^{n+\frac{1}{2}}) \tag{2.25}$$

$$\zeta_{ij}^{n+\frac{3}{2}} = f_3(U^{n+1}, V^{n+1}, \zeta^{n+\frac{1}{2}}) \tag{2.26}$$

facilitate the organisation of a simple integration algorithm subject to a numerical stability limit of the Δt value used. That limit is given by the known CFL criterion

$$\frac{c\Delta t}{\sqrt{(2\Delta x)}} < 1 \qquad (c = \sqrt{(gh)}) \tag{2.27}$$

From the physical aspect this inequality means that the Δt used must be less than the time needed for any perturbation to cover the extent of a mesh. This limit results in greater computer time than that required by an implicit scheme. The Δt used must of course be at least one or two orders of magnitude less than the characteristic time scale for the long wave to develop, so that no information is lost during the integration in time. For example, in the case of a tidal wave with $T = 43{,}000$ s, the optimal Δt would be $\Delta t = 430/4300$ s. Although the advent of more powerful computers makes this problem of secondary importance, implicit schemes have been investigated also for the integration of that numerical model. The most economic ones are based on the ADI technique, solving successively along the Ox and Oy directions implicitly and thus involving a large number of small algebraic systems instead of one large system in all the field unknowns.

The finite difference forms according to the present explicit scheme are:

$$\begin{aligned} U_{ij}^{n+1} = U_{ij}^{n} &- \frac{\Delta t}{8\Delta x}\left[(U_{i+1j}^{n} + U_{ij}^{n})^2 - (U_{ij}^{n} + U_{i-1j}^{n})^2\right] \\ &- \frac{\Delta t}{2\Delta y}\bar{V}_{ij}^{n}(U_{ij+1}^{n} - U_{ij-1}^{n}) - \frac{g\Delta t}{\Delta x}(\zeta_{ij}^{n+\frac{1}{2}}) - \zeta_{i-1}^{n+\frac{1}{2}} \\ &- \frac{2gU_{ij}^{n}\sqrt{(U_{ij}^{n2} + \bar{V}_{ij}^{n2})}}{C^2(h_{ij} + h_{i-1j})} + f\bar{V}_{ij}^{n} \end{aligned} \tag{2.28}$$

$$\begin{aligned} V_{ij}^{n+1} = V_{ij}^{n} &- \frac{\Delta t}{8\Delta y}\left[(V_{ij}^{n} + V_{ij+1}^{n})^2 - (V_{ij}^{n} + V_{ij-1}^{n})^2\right] \\ &- \frac{\Delta t}{2\Delta x}\bar{U}_{ij}^{n}(V_{i+1j}^{n} - V_{i-1j}^{n}) - \frac{g\Delta t}{\Delta y}(\zeta_{ij}^{n+\frac{1}{2}} - \zeta_{ij-1}^{n+\frac{1}{2}}) \\ &- \frac{2gV_{ij}^{n}\sqrt{(\bar{U}_{ij}^{n2} + V_{ij}^{n2})}}{C^2(h_{ij} + h_{ij-1})} - f\bar{U}_{ij}^{n} \end{aligned} \tag{2.29}$$

$$\begin{aligned} \zeta_{ij}^{n+\frac{3}{2}} = \zeta_{ij}^{n+\frac{1}{2}} &- \frac{\Delta t}{2\Delta x}(U_{i+1j}^{n+1}(h_{ij} + h_{i+1j}) - U_{ij}^{n+1}(h_{ij} + h_{i-1j})) \\ &- \frac{\Delta t}{2\Delta y}\left[V_{ij+1}^{n+1}(h_{ij} + h_{ij+1}) - V_{ij}^{n+1}(h_{ij} + h_{ij-1})\right] \\ &+ q_{ij}\Delta t \end{aligned} \tag{2.30}$$

where

$$\bar{V}_{ij}^{n} = (V_{ij}^{n} + V_{i-1j}^{n} + V_{ij+1}^{n} + V_{i-1j+1}^{n})/4$$

$$\bar{U}_{ij}^{n} = (U_{ij}^{n} + U_{ij-1}^{n} + U_{i+1j}^{n} + U_{i+1j-1}^{n})/4$$

The boundary conditions are treated very easily with the assumed approximation for the coastal boundaries and the use of a staggered grid. On the $\| Ox$ boundaries, $V_{ij}^{n} = 0$ and on the $\| Oy$ boundaries,

$U_{ij}^n = 0$. On the open sea boundaries either the total ζ_{ij}^n time series is known ($n = 1, 2, \ldots$) or the incident part of it ζ_i ($\zeta = \zeta_i + \zeta_r$) is known; for the radiated part ζ_r, Equation (2.11) is applied. On the boundary, the necessary velocity values have to be given or computed via relations based on the method of characteristics. A simple efficient approximation is the assumption of quasi-uniformity of the velocity across it, i.e. $\partial \mathbf{U}/\partial n = 0$.

The synthesis of a computer program analysis, Program 7, is straightforward. The most important points of the program are:

(1) The scanning of the field. The field is swept in the x direction (index i) for successive values of the ordinate y (index j). The leftmost and rightmost limits of the field for various j's are defined by the integer arrays IS(J), IE(J). An illustration of the procedure is included in Fig. 2.4.

(2) The special meshes are characterised by integer indices describing their properties. These are the meshes not belonging to the interior of the field but having one or more sides along a coastal or open sea boundary (dry meshes). This last type of mesh is used for the islands in multipli-connected flow domains. The characterisation and enumeration of the special meshes is illustrated in Fig. 2.4.

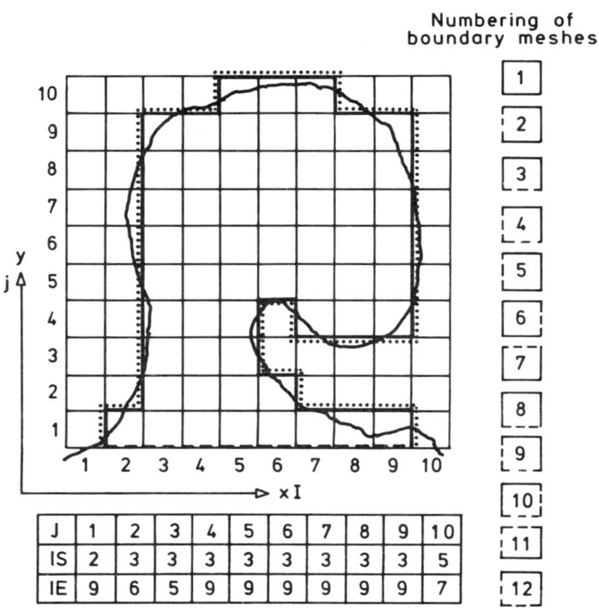

J	1	2	3	4	5	6	7	8	9	10
IS	2	3	3	3	3	3	3	3	3	5
IE	9	6	5	9	9	9	9	9	9	7

Fig. 2.4 Morphology of coastal boundaries and typical boundary meshes

(3) The U_{ij}^n, V_{ij}^n values are described by the U(I, J), V(I, J) arrays and the U_{ij}^{n+1}, V_{ij}^{n+1} by the UN(I, J), VN(I, J) arrays. In each time step the computed UN, VN values are stored as U, V arrays after their computation.
(4) After the computation of the velocity components on the mesh sides their mean values referring to the mesh center $(U_{ij} + U_{i+1j})/2$, $(V_{ij} + V_{ij+1})/2$ are computed. These are the plotted velocity values.
(5) The model computes a non-steady flow, evolving continuously in time due to a variable flow forcing factor (tidal flow). The same model can be used with a cold start (no flow) for the initial condition and can describe under a steady forcing factor the transient situation for a steady flow final state. It is a time-marching type technique. In each time-step the total kinetic energy in the flow domain is computed from the sum

$$E_{\text{kin}}^n = \sum_i \sum_j ((U_{ij}^n + U_{i+1j}^n)^2 + (V_{ij}^n + V_{ij+1}^n)^2) h_{ij} \Delta x \Delta y / 8 \tag{2.31}$$

The steady state is reached when the ratio $|E^{n+1} - E^n|/E^{n+1}$ becomes less than a test convergence value (10^{-3}).

PROGRAM 7: 2-D NON-LINEAR LONG-WAVE CIRCULATION MODEL

```
5REM 2-D NON LINEAR LONG WAVES MODEL
10DIMU(20,20),UN(20,20),V(20,20),VN(20,20),H(20
,20),Z(20,20),IS(20),IE(20),IB(50),JB(50),NB(50)
20READDT,DX,CF,F,IM,JM,KB,NM,AMPL,PER
30DATA...
40FORJ=1TOJM-1:READIS(J),IE(J):NEXTJ
50DATA....
60FORJ=1TOJM-1:FORI=IS(J)-1TOIE(J)+1:READH(I,J)
:NEXTI:NEXTJ
70DATA...
80FORI=IS(J)-1TOIE(J)+1
90FORK=1TOKB:READIB(K),JB(K),NB(K):NEXTK
100DATA...
110N=0:T=0:EK=0
120N=N+1:T=T+DT
130FORJ=2TOJM-2:FORI=IS(J)TOIE(J)
140Z(I,J)=Z(I,J)-DT/2/DX*(U(I+1,J)*(H(I,J)+H(I+1
,J))-U(I,J)*(H(I,J)+H(I-1,J))+V(I,J+1)*(H(I,J+1)+H
(I,J))-V(I,J)*(H(I,J)+H(I,J-1)))
145NEXTI:NEXTJ
150FORJ=2TOJM-2 :FORI=IS(J)+1TOIE(J)
160VV=(V(I,J)+V(I-1,J)+V(I,J+1)+V(I-1,J+1))/4:HM
=(H(I,J)+H(I-1,J))/2
170UN(I,J)=U(I,J)-DT*(((U(I,J)+U(I+1,J))^2-(U(I,
```

```
J)+U(I-1,J))^2)/8/DX+VV*(U(I,J+1)-U(I,J-1))/2/DX+9
.81*(Z(I,J)-Z(I-1,J))/DX-F*VV+CF*U(I,J)*SQR(VV^2+U
(I,J)^2)/HM)
  180NEXTI:NEXTJ
  190FORJ=3TOJM-2:FORI=IS(J)TOIE(J)
  200UU=(U(I,J)+U(I+1,J)+U(I,J-1)+U(I+1,J-1))/4:HM
=(H(I,J)+H(I,J-1))/2
  210VN(I,J)=V(I,J)-DT*(((V(I,J+1)+V(I,J))^2-(V(I,
J)+V(I,J-1))^2)/8/DX+UU*(V(I+1,J)-V(I-1,J))/2/DX+9
.81*(Z(I,J)-Z(I,J-1))/DX+F*UU+CF*V(I,J)*SQR(UU^2+V
(I,J)^2)/HM)
  220NEXTI :NEXTJ
  230FORK=1TOKB:I=IB(K):J=JB(K)
  3400N NB(K)-1 GOTO 350,360,370,380,390,400,410,4
20,430,440,450
  350UN(I,J)=0:GOTO465
  360VN(I,J)=0:GOTO465
  370UN(I,J)=0:VN(I,J)=0:GOTO465
  380UN(I,J)=UN(I+1,J):VN(I,J)=VN(I+1,J):GOTO460
  390UN(I+1,J)=UN(I,J):VN(I,J)=VN(I-1,J):GOTO460
  400VN(I,J+1)=VN(I,J):UN(I,J)=UN(I,J-1):GOTO460
  410UN(I,J)=UN(I,J+1):VN(I,J)=VN(I,J+1):GOTO460
  420UN(I,J)=UN(I+1,J):VN(I,J)=0:GOTO460
  430UN(I+1,J)=UN(I,J):VN(I,J)=0:GOTO460
  440VN(I,J+1)=VN(I,J):UN(I,J)=0:GOTO460
  450VN(I,J)=VN(I,J+1):UN(I,J)=0:GOTO460
  460Z(I,J)=AMPL*SIN(2*PI*T/PER)
  465NEXTK
  470FORJ=1TOJM:FORI=1TOIM:U(I,J)=UN(I,J):V(I,J)=V
N(I,J):NEXTI:NEXTJ
  480IFN/2<>INT(N/2) THEN GOTO120
  490EK=0:FORJ=2TOJM-2:FORI=IS(J)TOIE(J):EK=EK+((U
(I,J)+U(I+1,J))^2+(V(I,J)+V(I,J+1))^2)*H(I,J)/8 :N
EXTI:NEXTJ
  500PRINTN,EK
  510PRINT"UU":FORJ=JM-2 TO 2 STEP-1:FORI=2TOIM-2:
PRINT(U(I,J)+U(I+1,J))/2;:NEXTI:PRINT:NEXTJ
  520PRINT"VV":FORJ=JM-2 TO 2 STEP-1:FORI=2TOIM-2:
PRINT(V(I,J)+V(I,J+1))/2;:NEXTI:PRINT:NEXTJ
  530PRINT"ZZ" :FORJ=JM-2 TO 2 STEP-1:FORI=2TOIM-2
:PRINTZ(I,J);:NEXTI :PRINT:NEXTJ
  540IFN<NM THEN GOTO120
  550END
```

Description of main variables:

DT, DX	= discretisation in time and space (square mesh) steps
CF	= bed friction coefficient (g/C^2)
F	= Coriolis coefficient
IM, JM, NM	= max values of i, j, n indices in x, y, t directions

KB	= number of special meshes
AMPL	= long wave amplitude on the open sea boundary (in the program the ζ (total) values are prescribed)
PER	= period of long wave
IS(J), IE(J)	= lateral domain boundaries
H(I, J)	= water depths at mesh centers
IB, JB, NB	= abscissa, ordinate, and index characterising one of the KB special meshes (NB takes the value from 2 to 12).

The above program has a general validity. It can describe the tidal 2-D horizontal flow in any coastal basin with one open-sea boundary on which the total free surface elevation ζ has a prescribed sinusoidal variation. If on that boundary the incident component ζ_i is prescribed, then the open sea boundary is modified and the ζ_r component is computed via (2.11). An illustration of that procedure will be given in a subsequent program.

The present application refers to the circulation generated by a long wave in a semi-enclosed coastal basin of geometry given in Fig. 2.5.

The main data are: DT = 240 s, DX = 4000 m, CF = 0.01, F = 0.0001, IM = 9, JM = 9, KB = 11, NM = 200, AMPL = 1 m, PER = 4800 s.

IS	2	2	2	2	2	2	2				
IE	7	7	7	7	7	7	7				
IB	2	3	4	5	6	7	2	2	2	2	2
JB	2	2	2	2	2	2	3	4	5	6	7
NB	12	8	8	8	8	8	2	2	2	2	2

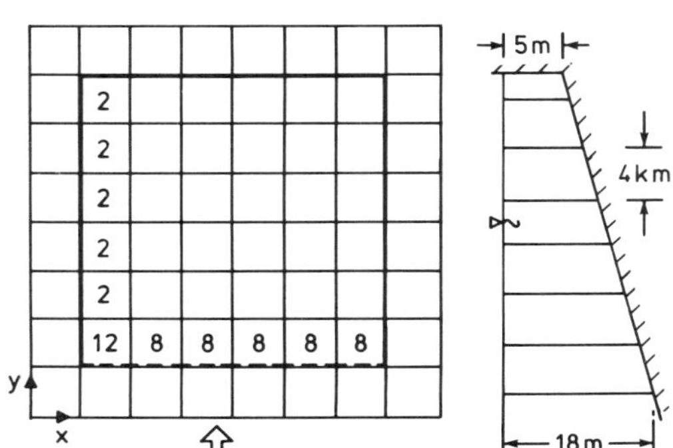

Fig. 2.5 Flow domain and discretisation for long-wave generated 2-D non linear model

The evolution of the free surface during a wave period ($t - t + T$) along two main directions of the flow domain is illustrated in Fig. 2.6. The influence of the Coriolis effect is shown from the surface slope along the section *b–b*. This slope produces the velocity component in the x direction.

The differences of the velocity hodographs (tidal velocity ellipses)

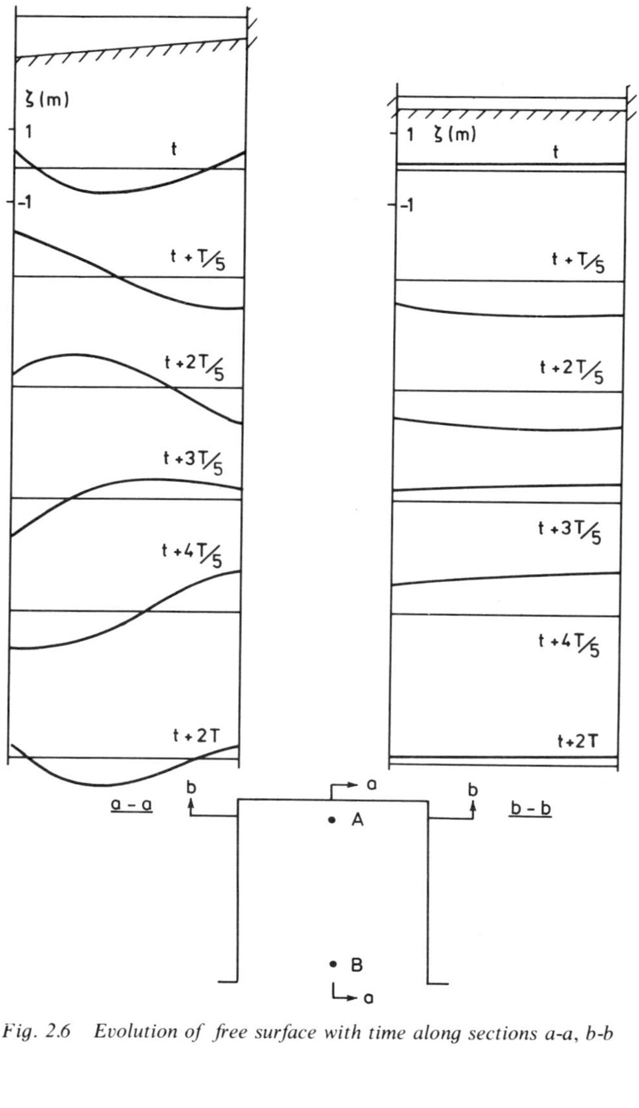

Fig. 2.6 Evolution of free surface with time along sections a-a, b-b

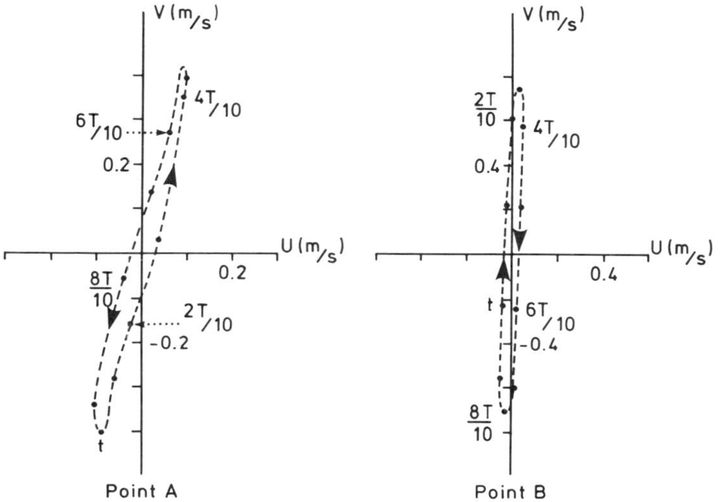

Fig. 2.7 Velocity hodographs at points A and B

at the points A and B, shown in Fig. 2.7, both with respect to the direction of rotation and the velocity magnitude, are due to the Coriolis effect and the depth difference.

2.2.3 1-D circulation model in a channel of varying cross-section

In some estuaries the flow domain has a length dimension much greater than the width. In that case the model can be further simplified by integration and averaging over the width dimension.

The velocities developing mainly along the longitudinal dimension can refer to the whole cross-section of the channel. The cross-section of the 1-D channel are approximated by orthogonal parallelograms in this version of the model. These may be of varying depth and width along the longitudinal axis Ox. The long-wave model (actually the generalised St. Venant model) is written in terms of the unknown functions of discharge $Q(x, t)$ and the water surface elevation $\zeta(x, t)$ measured from the horizontal SWL. The equilibrium of force and mass continuity equations, according to the notation of Fig. 2.8, are written:

$$\frac{\partial \zeta}{\partial t} + \frac{1}{B}\frac{\partial Q}{\partial x} = 0 \tag{2.32}$$

$$\frac{\partial Q}{\partial t} + \frac{\partial Q^2/A}{\partial x} = -gA\frac{\partial \zeta}{\partial x} - gAS_f \tag{2.33}$$

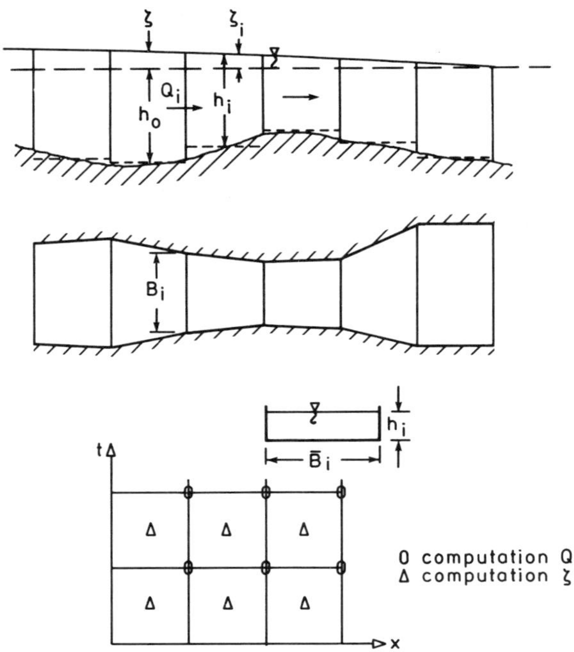

Fig. 2.8 1-D long-wave model. Flow domain discretisation and computation points in the x-t plane

where S_f the slope of the energy line, expressed according to the Chézy law, by the equation

$$S_f = \frac{U^2}{C^2 R} \qquad \left(R = \frac{A}{P} = \text{hydraulic radius}\right) \tag{2.34}$$

In (2.34) a local energy loss term, in abrupt channel enlargements, can be included, expressed as

$$\Delta h = (\Delta U)^2 / 2g \tag{2.35}$$

The numerical solution scheme to be presented is based on finite differences. An explicit scheme on a staggered grid is used. Figure 2.8 illustrates the location of the computation of Q, ζ functions on the x, t plane.

The finite difference form of Equations (2.32), (2.33), is

$$\frac{\zeta_i^{n+\frac{1}{2}} - \zeta_i^{n-\frac{1}{2}}}{\Delta t} = \frac{-2}{B_i + B_{i+1}} \cdot \frac{Q_{i+1}^n - Q_i^n}{\Delta x} \tag{2.36}$$

$$\frac{Q_i^{n+1}-Q_i^n}{\Delta t}=-\frac{(Q_{i+1}^n)^2/A_{i+1}-(Q_{i-1}^n)^2/A_{i-1}}{2\Delta x}-gA_i\frac{\zeta_i^{n+\frac{1}{2}}-\zeta_{i-1}^{n+\frac{1}{2}}}{\Delta x}$$

$$-gA_i\left[\frac{(U_i^n)^2}{C^2R_i}+\frac{(|U_{i+1}^n|-|U_{i-1}^n|)^2}{2g2\Delta x}\operatorname{sign}U_i^n\right] \qquad (2.37)$$

The last term in (2.37) describing the local losses is valid if $U_{i+1} < U_{i-1}$.

The boundary conditions on the first and last section of channel reaches depend on the physical processes there. On the first upstream reach usually the ζ or ζ_i variation with time is prescribed. In the first case that value is used directly in the equilibrium equation. In the second case a numerical solution of (1.33) has to be done for the computation of updated ζ_r values and ζ is expressed as the sum of $\zeta_r+\zeta_i$. If the incident perturbation is known on the first and second sections, for example $\zeta_i=\zeta_{i_0}\sin[(2\pi t/T)]$, then

$$\zeta_{i_1}^n=\zeta_{0i}\sin[2\pi(n-1)\Delta t/T] \qquad (2.38)$$

$$\zeta_{i_2}^n=\zeta_{0i}\sin[2\pi(n-1)\Delta t/T-\Delta x/L] \qquad (2.39)$$

where $L=T\sqrt{(gh)}$. Equation (2.39) is a direct consequence of the long progressive wave theory. Based on (2.38) and (2.39) the ζ_{r_1},ζ_{r_2} values are computed as the differences

$$\zeta_{r_1}^n=\zeta_1^n-\zeta_{i_1}^n \qquad (2.40)$$

$$\zeta_{r_2}^n=\zeta_2^n-\zeta_{i_2}^n \qquad (2.41)$$

These $\zeta_{r_1}^n,\zeta_{r_2}^n$ values, available at time level n, are used in an explicit backward difference scheme for the integration of (1.33) on the open sea boundary,

$$\zeta_{r_1}^{n+1}=\zeta_{r_1}^n+\frac{\Delta t}{\Delta x}C(\zeta_{r_2}^n-\zeta_{r_1}^n) \qquad (2.42)$$

The required ζ_1^{n+1} value on the boundary is the sum

$$\zeta_1^{n+1}=\zeta_{i_1}^{n+1}+\zeta_{r_1}^{n+1}=\zeta_0\sin(2\pi n\Delta t)/T+\zeta_{r_1}^{n+1} \qquad (2.43)$$

At the downstream end if there is a coastal boundary the full reflection condition is expressed as $Q=0$. If there is an open sea boundary the free transmission condition corresponding to (2.10) becomes,

$$Q_{i_{\max}}^{n+1}=\zeta_{i_{\max}-1}^{n+\frac{1}{2}}\sqrt{(9.81B_{i_{\max}}\cdot A_{i_{\max}})} \qquad (2.44)$$

Any other form of end conditions can be investigated. Program 8, in BASIC, refers to a flow domain with a given ζ upstream and downstream condition, either of full reflection or free transmission.

PROGRAM 8: 1-D NON-LINEAR TIDAL CIRCULATION MODEL

```
    5REM 1-D NONLINEAR TIDAL CIRCULATION MODEL
   10DIM Q(50),QN(50),H(50),Z(50),R(50),B(50),A(50
),HO(50)
   20READ IM,DT,DX,C,PR,ZO,NM,BK
   25DATA...
   30FORI=1TOIM:READB(I):NEXTI
   35DATA...
   40FORI=1TOIM:READHO(I):H(I)=HO(I):NEXTI
   45DATA...
   50N=0:T=0
   60N=N+1:T=T+DT
   70FORI=2TOIM-1:H(I)=HO(I)+(Z(I)+Z(I-1))/2:NEXTI
  :H(1)=HO(1)+Z(1):H(IM)=HO(IM)+Z(IM-1)
   80FORI=1TOIM:A(I)=B(I)*H(I):R(I)=A(I)/(B(I)+2*H
(I)):NEXTI
   90FORI=2TOIM-1:Z(I)=Z(I)-DT/DX*(Q(I+1)-Q(I))/(B
(I)+B(I+1))*2:NEXTI
  100Z(1)=ZO*SIN(2*PI*T/PR)
  110FORI=2TOIM-1:VV=0
  120IFB(I+1)>B(I-1)ANDQ(I)>0 THEN GOTO150
  130IFB(I+1)<B(I-1) ANDQ(I)<0 THEN GOTO150
  140GOTO160
  150VV=(ABS(Q(I+1)/A(I+1))-ABS(Q(I-1)/A(I-1)))^2/
4./9.8/DX
  160QN(I)=Q(I)-DT*(Q(I+1)^2/A(I+1)-Q(I-1)^2/A(I-1
))/2/DX-DT*9.8*A(I)*(Z(I)-Z(I-1))/DX-DT*9.8*A(I)*(
(Q(I)/A(I))^2/C^2/R(I)+VV)*SGN(Q(I)):NEXTI
  170QN(1)=QN(2)
  180ON BK GOTO190,200
  190QN(IM)=Z(IM-1)*SQR(9.8*B(IM)*A(IM)):GOTO210
  200QN(IM)=0
  210FORI=1TOIM:Q(I)=QN(I):NEXTI
  220IFN/20<>INT(N/20) THEN GOTO60
  230PRINT:PRINTN:FORI=1TOIM:PRINTZ(I);:NEXTI:PRIN
T:FORI=1TOIM:PRINTQ(I)/A(I);:NEXTI:PRINT
  240IFN<NM THEN GOTO60
  250END
>
```

Description of main variables:

DT, DX	= time and space discretisation steps
IM	= number of cross-sections in the field discretisation
C	= Chézy bed friction coefficient
PR	= long-wave period
Z0	= long-wave amplitude
NM	= number of time steps for the numerical solution

BK = parameter defining the type of downstream boundary condition (BK = 1 → free transmission BK = 2 → full reflection)
B(I) = width of cross-section #i
H0(I) = initial depth at section #i.

The application refers to a flow domain illustrated in Fig. 2.9. It is a long channel of length = 2 km, of depth = 10 m and width varying from 30 to 10 m. At the upstream end the free surface varies sinusoidally with period $T = 500$ s and amplitude 1 m.

The evolution of a free surface and the velocity along the channel is required. A free transmission downstream boundary condition (connection of the channel end to a large water body) is assumed. The data are: IM = 21, DT = 5 s, DX = 100 m, C = 50 $m^{\frac{1}{2}}$/s, PR = 500 s, Z0 = 1 m, NM = 1000, BK = 1, H0(I) = 10 m, B(I) varying linearly from 30 to 10 m. Figure 2.10 contains the free surface profiles along the channel during one wave cycle after the establishment of periodic conditions in the channel. Due to the type of downstream boundary condition, this is achieved only after the 2nd wave cycle, starting with no flow initial conditions ($Q_i = 0$, $\zeta_i = 0$, for $t = 0$).

2.2.4 Linearised model for long-wave induced circulation

A realistic simplification of the mathematical model for long-wave induced circulation is introduced by linearisation (disregarding non-linear convective and friction terms) and leads to a practical and efficient form. The formulation starts from Equations (2.21) and (2.22), dropping the convective terms. Solving successively the two equations for the ζ function in one dimension, it is found:

$$\frac{\partial U}{\partial t} = -g\frac{\partial \zeta}{\partial x} - g\frac{U|U|}{hC^2} \tag{2.45}$$

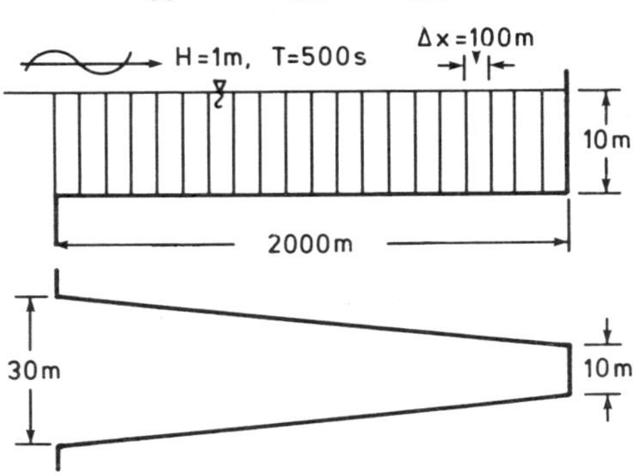

Fig. 2.9 Flow domain discretisation for 1-D long-wave model

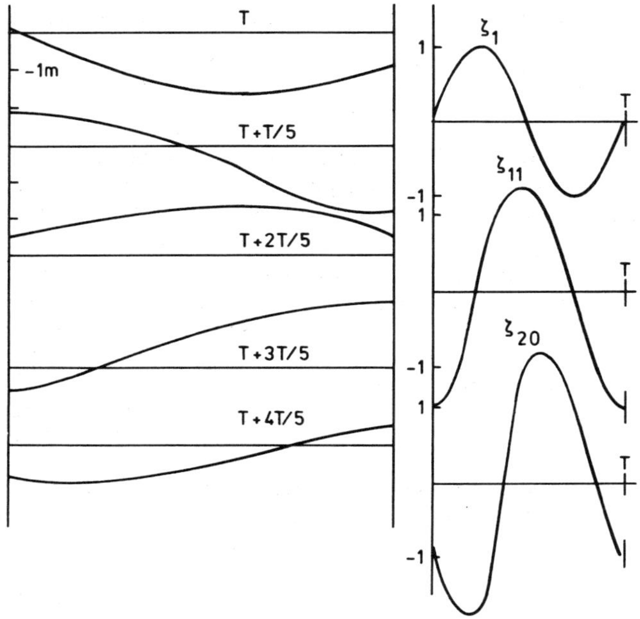

Fig. 2.10 Surface time history during a wave period

$$\frac{\partial \zeta}{\partial t} + \frac{\partial hU}{\partial x} = 0 \tag{2.46}$$

$$\frac{\partial^2 \zeta}{\partial t^2} = \frac{\partial}{\partial x}\left(gh\frac{\partial \zeta}{\partial x}\right) - \frac{2g|U|}{hC^2} \cdot \frac{\partial \zeta}{\partial t} \tag{2.47}$$

The bed friction term in (2.47) has the form of the product of the time derivative by a term that can be considered as a friction coefficient

$$k = \frac{2g|U|}{hC^2} \qquad \text{where } C = 18 \log\left(\frac{12h}{k_N}\right) \tag{2.48}$$

and $|U|$ is a measure of the velocity. The application of long-wave theory permits the correlation of $|U|$ to the wave and the depth ($u = u_0 \sin(2pt/T)$ where $u_0 = H/2\sqrt{(g/h)}$). The approximation of $|U|$ with its root mean value $\sqrt{(\bar{u}^2)}$ leads to an expression for k,

$$k = \frac{2g^{3/2}H}{3C^2h^{3/2}} \tag{2.49}$$

Equation (2.47) generalised in two dimensions is written

$$\frac{\partial^2 \zeta}{\partial t^2} = \frac{\partial}{\partial x}\left(gh\frac{\partial \zeta}{\partial x}\right) + \frac{\partial}{\partial y}\left(gh\frac{\partial \zeta}{\partial y}\right) - k\frac{\partial \zeta}{\partial t} \qquad (2.50)$$

This is a 2nd order hyperbolic equation completed by a friction term (generalised telegraphy equation). It describes the variation only of ζ in time and space. If the velocity magnitudes are necessary at various locations of the flow domain those can be computed from the known ζ and grad ζ values by time integration of equations having the form of (2.45). Equation (2.50) does not contain the Coriolis term. Simple mathematical manipulation shows that when that term is important the right hand side of (2.50) has to be completed by a term $-f^2 \cdot \zeta$ where $f = 2\Omega \sin \varphi$.

The field Equation (2.50) is completed by boundary conditions of two types:

(1) At the perimeter of the coastal boundary where full reflection of the long wave takes place

$$\frac{\partial \zeta}{\partial n} = 0 \qquad (2.51)$$

(2) At the open sea boundary, the condition of free radiation of the ζ_r part is

$$\frac{\partial \zeta_r}{\partial t} + \sqrt{(gh)}\frac{\partial \zeta_r}{\partial n} = 0 \qquad (2.52)$$

where n is the unit outward vector normal to that boundary. The incident component ζ_i is known. The numerical solution of (2.50) is based on an explicit centered finite difference scheme. On the i, j nodes of the orthogonal mesh discretising the flow domain, the ζ_{ij}^n values are computed at each time level n ($t_n = n \cdot Dt$). The h_{ij} values referring to the node i, j are used. Equation (2.50) takes the form,

$$\begin{aligned}\zeta_{ij}^{n+1} = 2\zeta_{ij}^n - \zeta_{ij}^{n-1} + \frac{\Delta t^2}{2\Delta x^2} g[(h_{i+1j} + h_{ij})(\zeta_{i+1j}^n - \zeta_{ij}^n) \\ - (h_{ij} + h_{i-1j})(\zeta_{ij}^n - \zeta_{i-1j}^n) + (h_{ij+1} + h_{ij})(\zeta_{ij+1}^n - \zeta_{ij}^n) \\ - (h_{ij} + h_{ij-1})(\zeta_{ij}^n - \zeta_{ij-1}^n)] - k\Delta t(\zeta_{ij}^n - \zeta_{ij}^{n-1})\end{aligned} \qquad (2.53)$$

On the coastal boundary not parallel to the Ox, Oy axes and not coinciding with mesh sides, the approximation of the $\partial\zeta/\partial n$ derivative introduces an external node to the computations where the ζ value,

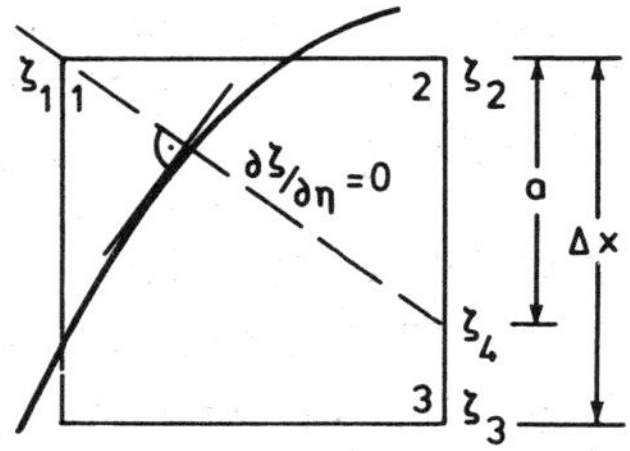

Fig. 2.11 Coastal boundary node notations for 2-D linear long-wave model

according to the notation of Fig. 2.11, is approximated by

$$\zeta_1 = \zeta_4, \qquad \zeta_4 = \zeta_2 + (\zeta_3 - \zeta_2)a/\Delta x \qquad (2.54)$$

where the ζ_2, ζ_3 are known internal values and a is a known line segment.

A linear long-wave model based on the forgoing numerical analysis is programmed in BASIC, Program 9. The program refers to a flow domain bounded laterally by a coast of arbitrary geometry and presenting a lower open-sea boundary parallel to Ox. The bathymetry is also variable.

PROGRAM 9: 2-D LINEAR TIDAL CIRCULATION MODEL

```
>L.
    5REM 2-D LINEAR TIDAL CIRCULATION MODEL
   10DIMH(20,20),Z1(20,20),Z(20,20),Z0(20,20),IS(2
0),IE(20),I1(50),I2(50),I3(50),J1(50),J2(50),J3(50
),EL(50)
   20READ DT,DX,FR,A0,CF,IM,JM,NM,BR
   30DATA...
   40FORJ=1TOJM-1:READIS(J),IE(J):NEXTJ
   50DATA...
   60FORK=1TOBR:READI1(K),J1(K),I2(K),J2(K),I3(K),
J3(K),EL(K):NEXTK
   70DATA...
   80FORJ=1TOJM-1:FORI=IS(J)TOIE(J):H(I,J)=30:NEXT
I:NEXTJ
  110T=0:N=0
  120T=T+DT:N=N+1
  130FORJ=2TOJM-1:FORI=IS(J)TOIE(J):Z1(I,J)=2*Z(I,
J)-Z0(I,J)+DT^2/DX^2/2*9.81*((H(I+1,J)+H(I,J))*(Z(
I+1,J)-Z(I,J))-(H(I,J)+H(I-1,J))*(Z(I,J)-Z(I-1,J))
+(H(I,J)+H(I,J+1))*(Z(I,J+1)-Z(I,J))-(H(I,J)+H(I,J
-1))*(Z(I,J)-Z(I,J-1)))
  140Z1(I,J)=Z1(I,J)-CF*DT*(Z(I,J)-Z0(I,J)):NEXTI:
```

```
NEXTJ
  160FORK=1TOBR:I1=I1(K):I2=I2(K):I3=I3(K):J1=J1(K
):J2=J2(K):J3=J3(K)
  170Z1(I1,J1)=Z1(I2,J2)+(Z1(I3,J3)-Z1(I2,J2))*EL(
K):NEXTK
  180FORI=IS(1)TOIE(1):C=SQR(9.81*H(I,1)):L=C*PR
  190IF(T-DT)>DX/C THEN GOTO210
  200Z1=0:GOTO220
  210Z1=Z(I,1)-A0*SIN(2*PI*(T-DT)/PR):Z2=Z(I,2)-A0
*SIN(2*PI*(T-DT)/PR-DX/L):Z1=Z1+DT/DX*C*(Z2-Z1)
  220Z1(I,1)=Z1+A0*SIN(2*PI*T/PR):NEXTI
  230FORJ=1TOJM:FORI=1TOIM:Z0(I,J)=Z(I,J):Z(I,J)=Z
1(I,J):NEXTI:NEXTJ
  240IFN/10<>INT(N/10) THEN GOTO120
  250PRINT:PRINTT:FORJ=JM TO 1 STEP-1:FORI=1TOIM:P
RINTZ(I,J);:NEXTI:PRINT:NEXTJ
  260IFN<NM THEN GOTO120
  270END
```

Description of main variables:

DT, DX = time and space discretisation steps
PR = period of incident wave
A0 = amplitude of incident wave
CF = bed friction coefficient (k)
IM, JM = maximum values of i, j indices along Ox, Oy
NM = time steps of computation
BK = number of coastal nodes on exterior and interior (islands) coastal boundaries. The indices I_1, J_1 (coordinates of the node) and I_2, J_2, I_3, J_3 (coordinates of related internal nodes) according to Fig. 2.11 and Equation (2.54)
EL = the corresponding ratio a/Dx in (2.54)
IS, IE = coupled values of i index for each j, indicating lateral limits (leftmost and rightmost limits respectively) of the flow domain.

The application described below is intended to demonstrate the variation of wave height between the entrance and innermost boundary of a coastal basin. The flow domain and its discretisation are given in Fig. 2.12. The data for that application is: DT = 30 s, DX = 1000 m, PR = 2000 s, CF = $3 \cdot 10^{-4}$ (corresponding to Chézy C = 20 m$^{\frac{1}{2}}$/s), IM = 9, JM = 9, NM = 300, BK = 15, A0 = 1 m, H(I, J) = 30 m. The rest of the data are given in Tables 2.1 and 2.2. Figure 2.13 contains the free surface tide curves at the bay entrance and the inner coastal boundary.

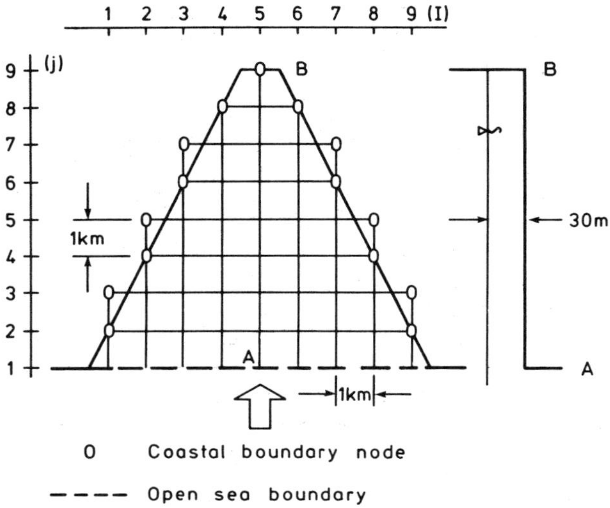

Fig. 2.12 Coastal domain of 2-D linear long-wave model application

Table 2.1

J	IS	IE
1	2	8
2	2	8
3	2	8
4	3	7
5	3	7
6	4	6
7	4	6
8	5	5

2.3 WIND GENERATED CIRCULATION

The wind generated circulation model describes the phenomenon under the same assumptions made in Section 2.1. So the general circulation equations (2.5) to (2.7) are valid in this case, too.

The circulation is forced by the shear stresses on the water surface exercised by the wind. They appear in the model in the form of free surface boundary conditions (2.13), (2.14). For small scale geophysical domains ($O[L] = 10^4$ m) the wind velocity may be assumed uniform and thus the components τ_{sx}, τ_{sy} are constant in

Table 2.2

K	I_1	J_1	I_2	J_2	I_3	J_3	EL
1	1	2	2	2	2	1	0.5
2	1	3	2	3	2	2	0.5
3	2	4	3	4	3	3	0.5
4	2	5	3	5	3	4	0.5
5	3	6	4	6	4	5	0.5
6	3	7	4	7	4	6	0.5
7	4	8	5	8	5	7	0.5
8	5	9	5	8	5	8	0
9	6	8	5	8	5	7	0.5
10	7	7	6	7	6	6	0.5
11	7	6	6	6	6	5	0.5
12	8	5	7	5	7	4	0.5
13	8	4	7	4	7	3	0.5
14	9	3	8	3	8	2	0.5
15	9	2	8	2	8	1	0.5

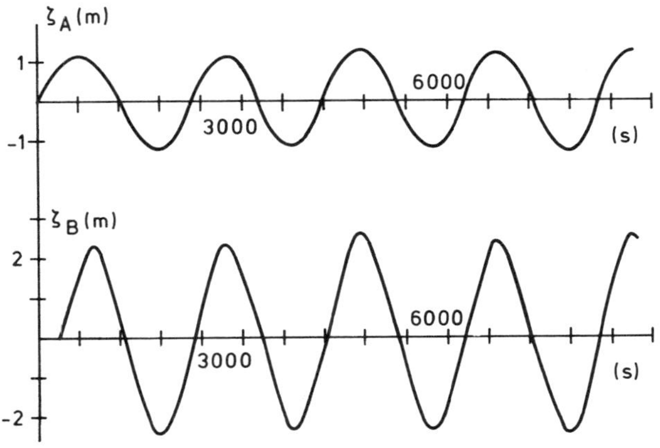

Fig. 2.13 Evolution of free surface at points A, B

space but variable in time. For larger geophysical domains this assumption is not realistic and the influence of wind nonuniformity has to be checked.

The wind generated waves are not incorporated in the model. Only the wind induced shear. The free surface is approximated by the mean level (with respect to the waves). The inclusion of the influence of waves in the wind generated circulation can be done implicitly in a parametric manner during the model calibration (determination of the surface friction coefficient and eddy viscosity distribution by comparison with the in situ measurements).

A solution of the general circulation model is not given in this book—only a comment on the eddy viscosity distribution over the vertical is considered worth mentioning. As the stronger velocity gradients appear near both the bed and the surface, the turbulence intensity and vertical momentum diffusion are related to both u_{b*} and $u_{s*} = \sqrt{(\tau_s/\rho)}$. The eddy viscosity distribution, deriving from higher order turbulence closure, simulates in an optimal way the wind generated profile (see Fig. 2.14). The maximum is at a distance $1/3h$ from the surface and its value is proportional to

$$\nu_{\max} \propto \lambda u_{s*} h, \qquad O[\lambda] = 0.1 \tag{2.55}$$

Figure 2.14 contains morphologies of current profiles in domains confined laterally by coastal boundaries. In such domains the current direction and intensity vary considerably along the depth. At the surface the current follows the wind direction (with declination to the right in the Northern hemisphere). Some distance below the surface, the direction is reversed due to return flows imposed by coastal boundaries.

These comments on the general morphology of the wind generated current profiles show that simple depth averaging of the general model (analogous to that for the long-wave induced circulation) has

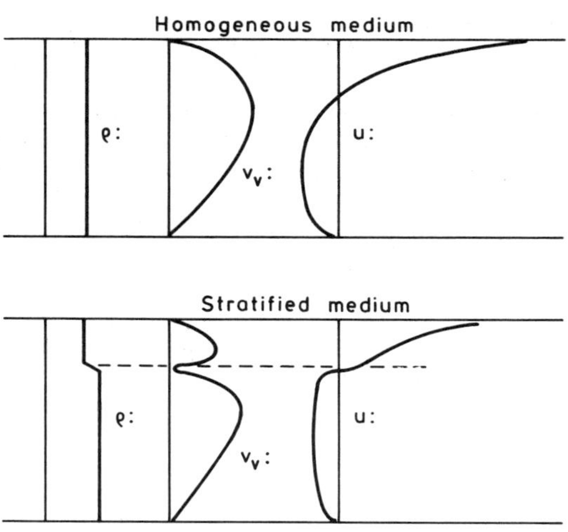

Fig. 2.14 Distribution of wind generated current in 1-D enclosed basin for homogeneous and stratified medium

to be done very carefully. The simple 2-D horizontal model used operationally in the past for wind generated circulation has the form:

$$\frac{\partial U}{\partial t} + U\frac{\partial U}{\partial x} + V\frac{\partial U}{\partial y} = -g\frac{\partial \zeta}{\partial x} + fV + \frac{\tau_{sx}}{\rho h} - \frac{\tau_{bx}}{\rho h} \tag{2.56}$$

$$\frac{\partial V}{\partial t} + U\frac{\partial V}{\partial x} + V\frac{\partial V}{\partial y} = -g\frac{\partial \zeta}{\partial y} - fU + \frac{\tau_{sy}}{\rho h} - \frac{\tau_{by}}{\rho h} \tag{2.57}$$

$$\frac{\partial \zeta}{\partial t} + \frac{\partial (Uh)}{\partial x} + \frac{\partial (Vh)}{\partial y} = 0 \tag{2.58}$$

where the bed friction terms are expressed by quadratic forms:

$$\frac{\tau_{bx}}{\rho} = kU\sqrt{(U^2 + V^2)}, \qquad \frac{\tau_{by}}{\rho} = kV\sqrt{(U^2 + V^2)} \tag{2.59}$$

The criticism that can be made of that model refers to:

(1) The depth average approximation of the convective terms:

$$\frac{1}{h}\int_{-h}^{0} u\frac{\partial u}{\partial x}\,\mathrm{d}z \simeq U\frac{\partial U}{\partial x} \tag{2.60}$$

As the $u(z), v(z)$ distributions are highly non-uniform (2.60) is inaccurate. The correction that can be made to (2.60) can take the form of an additional term known as horizontal momentum dispersion term.

(2) Equation (2.59) implies that when the depth-mean velocity components are zero, the friction terms are suppressed. From the typical morphology of wind generated current profile it can be concluded that even in the case of zero depth-mean velocities the near bed shear is not negligible and acts in the same direction as the surface shear. This means that a better approximation to τ_{bx}, τ_{by} than (2.59) has to be used.

A first upgrading of the model (2.56), (2.57) without solving in three-dimensional space can be made by adopting a certain distribution of current over the depth. Let us assume that the $u(z)$ form is:

$$u(z) = \alpha z^2 + \beta z + \gamma \tag{2.61}$$

The following conditions are used for computation of the undetermined coefficients α, β, γ:

(1) Free surface condition

$$\left.\frac{\partial u}{\partial z}\right|_{z=0} = \frac{\tau_s}{\rho \nu} \tag{2.62}$$

(2) Bed condition

$$\left. u \right|_{z=-h} = 0 \tag{2.63}$$

(3) Depth mean velocity definition

$$U = \frac{1}{h}\int_{-h}^{0} u \, \mathrm{d}z \tag{2.64}$$

The application of Equations (2.62) to (2.64) in (2.61) gives

$$u(z) = (\tfrac{3}{4}a - \tfrac{3}{2}U)\left[\left(\frac{z}{h}\right)^2 - 1\right] + a\left(\frac{z}{h} + 1\right) \tag{2.65}$$

where

$$a = \frac{\tau_s h}{\rho \nu}$$

The problem is transposed to the determination of the eddy viscosity ν (at the surface). In order to be consistent with the parabolic velocity distribution a constant eddy viscosity is assumed with mean value

$$\bar{\nu} = \lambda h \sqrt{\left(\frac{\tau_s}{\rho}\right)} \tag{2.66}$$

Turbulence models and laboratory measurements indicate that $O[\lambda] = 0.1$. For $\bar{\lambda} = 0.066$, it is found that

$$a = \frac{h\tau_s}{\rho \nu} = 16.6 \sqrt{\left(\frac{\tau_s}{\rho}\right)}$$

and the velocity distribution (2.65) is expressed in terms of known magnitudes and the mean depth $U(x, y)$. The substitution of (2.65) and (2.66) in the bed shear expression $\tau_b/\rho = \nu \, \partial u/\partial z$ gives for τ_b/ρ:

$$\frac{\tau_b}{\rho} = 0.18 \sqrt{\left(\frac{\tau_s}{\rho}\right)} U - 0.5 \frac{\tau_s}{\rho} \tag{2.67}$$

Finally, the use of (2.65) in the integral of the convective term $u \, \partial u/\partial x$ along the depth gives:

$$\frac{1}{h}\int_{-h}^{0} u \frac{\partial u}{\partial x} \, \mathrm{d}z = U \frac{\partial U}{\partial x} + \left(0.2U + \frac{a}{40}\right)\frac{\partial U}{\partial x} \tag{2.68}$$

The 2DH model, improved with respect to the horizontal momentum dispersion and the bed friction for wind generated circulation, becomes on the basis of (2.67) and (2.68):

$$\frac{\partial U}{\partial t}+U\frac{\partial U}{\partial x}+V\frac{\partial U}{\partial y}+\left(0.2U+\frac{a_x}{40}\right)\frac{\partial U}{\partial x}+\left(0.2V+\frac{a_y}{40}\right)\frac{\partial U}{\partial y}$$

$$=-g\frac{\partial \zeta}{\partial x}+fV+\frac{\tau_{sx}}{\rho h}-\left(0.18\frac{U}{h}\sqrt{\left(\frac{\tau_s}{\rho}\right)}-0.5\frac{\tau_{sx}}{\rho h}\right) \qquad (2.69)$$

$$\frac{\partial V}{\partial t}+U\frac{\partial V}{\partial x}+V\frac{\partial V}{\partial y}+\left(0.2U+\frac{a_x}{40}\right)\frac{\partial V}{\partial x}+\left(0.2V+\frac{a_y}{40}\right)\frac{\partial V}{\partial y}$$

$$=-g\frac{\partial \zeta}{\partial y}-fU+\frac{\tau_{sy}}{\rho h}-\left(0.18\frac{V}{h}\sqrt{\left(\frac{\tau_s}{\rho}\right)}-0.5\frac{\tau_{sy}}{\rho h}\right) \qquad (2.70)$$

The continuity equation keeps its form (2.58).

The coastal boundary conditions are the same as those for long-wave circulation. On the open sea boundary where there is no incidence perturbation waves coming from inside are freely radiated by application of condition (2.10).

Based on the same explicit finite difference scheme on a staggered grid as that used for the long-wave induced circulation, the computer program synthesized for the wind generated model is little different from the previous one. The only variation appears in the statements for the computation of UN, VN (the velocity components) and in the application of the open-sea boundary condition. By way of repetition a BASIC program is presented which refers to a flow domain with arbitrary coastal geometry, Program 10. The wind influence is assumed constant over the flow field. A time series of wind velocity components could be introduced in the case of wind varying with time. The initial condition is one of a cold start and the transient development of the hydrodynamic conditions to steady flow is followed. The printed results display U, V and ζ values at the mesh centers. Equation (2.65) permits the computation of the current pattern at any depth. If, for example, the free surface velocity components are required (for subsequent use in a surface advective diffusive pollutant transport model) they can be computed from (2.65)

$$u_{\text{surf}}=1.5U+a_x/4 \qquad (2.71)$$

$$v_{\text{surf}}=1.5V+a_y/4 \qquad (2.72)$$

PROGRAM 10: 2-D MODIFIED WIND GENERATED CIRCULATION MODEL

```
    5REM 2-D MODIFIED WIND GENERATED CIRCULATION M
ODEL
   10DIMU(20,20),UN(20,20),V(20,20),VN(20,20),H(20
,20),Z(20,20),IS(20),IE(20),IB(50),JB(50),NB(50)
   20READDT,DX,CS,WX,WY,F,IM,JM,KB,NM
   30DATA...
   40FORJ=1TOJM-1:READIS(J),IE(J):NEXTJ
   50DATA...
   60FORJ=1TOJM-1  :FORI=IS(J)-1TOIE(J)+1:READH(I,J
):NEXTI:NEXTJ
   70DATA...
   80TX=CS*WX*SQR(WX^2+WY^2):TY=CS*WY*SQR(WX^2+WY^
2):TS=SQR(TX^2+TY^2)
   90FORK=1 TO KB:READIB(K),JB(K),NB(K):NEXTK
  100DATA...
  110N=0:T=0:EK=0
  120N=N+1:T=T+DT
  130FORJ=2TOJM-2:FORI=IS(J)TOIE(J):Z(I,J)=Z(I,J)-
DT/2/DX*(U(I+1,J)*(H(I,J)+H(I+1,J))-U(I,J)*(H(I,J)
+H(I-1,J))+V(I,J+1)*(H(I,J+1)+H(I,J))-V(I,J)*(H(I,
J)+H(I,J-1))):NEXTI:NEXTJ
  140FORJ=2TOJM-2:FORI=IS(J)+1TOIE(J):VV=(V(I,J)+V
(I-1,J)+V(I,J+1)+V(I-1,J+1))/4:HM=(H(I,J)+H(I-1,J)
)/2
  150AD=(1.2*U(I,J)+.4*SGN(TX)*SQR(ABS(TX)))*(U(I+
1,J)-U(I-1,J))/2/DX+(1.2*VV+.4*SGN(TY)*SQR(ABS(TY)
))*(U(I,J+1)-U(I,J-1))/2/DX:BF=.18*U(I,J)/HM*SQR(T
S)-.5*TX/HM:UN(I,J)=U(I,J)-DT*(AD+9.81*(Z(I,J)-Z(I
-1,J))/DX-F*VV+BF-TX/HM):NEXTI:NEXTJ
  160FORJ=3TOJM-2:FORI=IS(J)TOIE(J):UU=(U(I,J)+U(I
+1,J)+U(I,J-1)+U(I+1,J-1))/4:HM=(H(I,J)+H(I,J-1))/
2
  170AD=(1.2*V(I,J)+.4*SGN(TY)*SQR(ABS(TY)))*(V(I,
J+1)-V(I,J-1))/2/DX+(1.2*UU+.4*SGN(TX)*SQR(ABS(TY)
))*(V(I+1,J)-V(I-1,J))/2/DX:BF=.18*V(I,J)/HM*SQR(T
S)-.5*TY/HM:VN(I,J)=V(I,J)-DT*(AD+9.81*(Z(I,J)-Z(I
,J-1))/DX+F*UU+BF-TY/HM):NEXTI:NEXTJ
  180FORK=1TOKB:I=IB(K):J=JB(K):ON NB(K)-1 GOTO 19
0,200,210,220,230,240,250,260,270,280,290
  190UN(I,J)=0:GOTO310
  200VN(I,J)=0:GOTO310
  210UN(I,J)=0:VN(I,J)=0:GOTO310
  220UN(I,J)=-Z(I,J)*SQR(9.81/H(I,J)):VN(I-1,J)=VN
(I,J):GOTO310
  230UN(I+1,J)=Z(I,J)*SQR(9.81/H(I,J)):VN(I+1,J)=V
N(I,J):GOTO310
  240VN(I,J+1)=Z(I,J)*SQR(9.81/H(I,J)):UN(I,J+1)=U
```

```
N(I,J):GOTO310
  250VN(I,J)=-Z(I,J)*SQR(9.81/H(I,J)):UN(I,J-1)=UN
(I,J):GOTO310
  260UN(I,J)=-Z(I,J)*SQR(9.81/H(I,J)):VN(I,J)=0:GO
TO310
  270UN(I+1,J)=Z(I,J)*SQR(9.81/H(I,J)):VN(I,J)=0:G
OTO310
  280UN(I,J)=0:VN(I,J+1)=Z(I,J)*SQR(9.81/H(I,J)):G
OTO310
  290UN(I,J)=0:VN(I,J)=-Z(I,J)*SQR(9.81/H(I,J))
  310NEXTK
  320FORJ=1TOJM:FORI=1TOIM:U(I,J)=UN(I,J):V(I,J)=V
N(I,J):NEXTI:NEXTJ
  330IFN/50<>INT(N/50) THEN GOTO120
  340KK=EK:EK=0:FORJ=2TOJM-2:FORI=IS(J)TOIE(J):EK=
EK+((U(I,J)+U(I+1,J))^2+(V(I,J)+V(I,J+1))^2)*H(I,J
)/8:NEXTI:NEXTJ
  350PRINTN,EK
  360IFABS(EK-KK)/EK>.0001 OR N<NM THEN GOTO120
  370PRINT"UU":FORJ=JM-2 TO 2 STEP-1:FORI=2TOIM-2:
PRINT(U(I,J)+U(I+1,J))/2;:NEXTI:PRINT:NEXTJ
  380PRINT"VV":FORJ=JM-2 TO 2 STEP-1:FORI=2TOIM-2:
PRINT(V(I,J)+V(I,J+1))/2;:NEXTI:PRINT:NEXTJ
  390PRINT"ZZ":FORJ=JM-2 TO 2 STEP-1:FORI=2TOIM-2:
PRINTZ(I,J);:NEXTI:PRINT:NEXTJ
  400END
```

Description of main variables:

DX, DT	= space and time discretisation steps
CS	= wind friction coefficient K
WX, WY	= wind velocity components along Ox, Oy
F	= Coriolis coefficient
IM, JM	= maximum values of I, J grid indices
KB	= number of boundary (coastal + open sea) meshes to be specially treated
NM	= maximum number of time steps
IS(J), IE(J)	= leftmost and rightmost values of mesh index i for various ordinates j
H(I, J)	= water depths at mesh centers
IB, JB, NB	= coordinate indices for boundary meshes and index denoting the type of boundary. For NB, the numbering as shown in Fig. 2.4 is used.

The application is a comparative presentation of the numerical solutions of the two 2DH models of wind generated circulation, the modified one and the classical one without correction of the $u(\partial u/\partial x)$, τ_b/ρ terms. The flow domain morphology and its discretisation by a

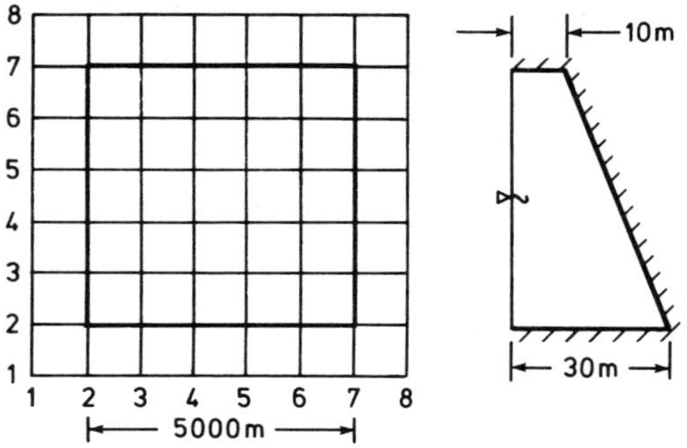

Fig. 2.15 Flow domain and discretisation for wind generated circulation model

square grid is given in Fig. 2.15. Program data: DT = 30 s, DX = 1000 m, CS = 0.000 005 (an exaggerated value as $O[k] = 1 \div 3 \times 10^{-6}$), WX = 10 m/s, WY = 10 m/s, IM = 8, JM = 8, KB = 9, NM = 2000.

IS	2	2	2	2	2	2	2		
IE	6	6	6	6	6	6	6		
IB	2	2	2	2	2	3	4	5	6
JB	2	3	4	5	6	2	2	2	2
NB	4	2	2	2	2	3	3	3	3

Figure 2.16 illustrates the development of the kinetic energy up to the establishment of steady flow conditions, the steady flow U, V fields and the surface profile along the diagonal oriented parallel to the wind. The difference between the two models in the kinetic energy development and the storm surge is obvious. Regarding the latter, it has to be said that the corrected model in the case of wind blowing normal to a coast (with small depth mean velocities) results in higher free surface gradients balancing both the free surface shear and the bed shear (in the same direction) and consequently to higher storm surges along the coast.

2.4 WAVE GENERATED CIRCULATION

Wave generated circulation describes the mean motion that is generated in coastal areas where wind-generated short waves refract,

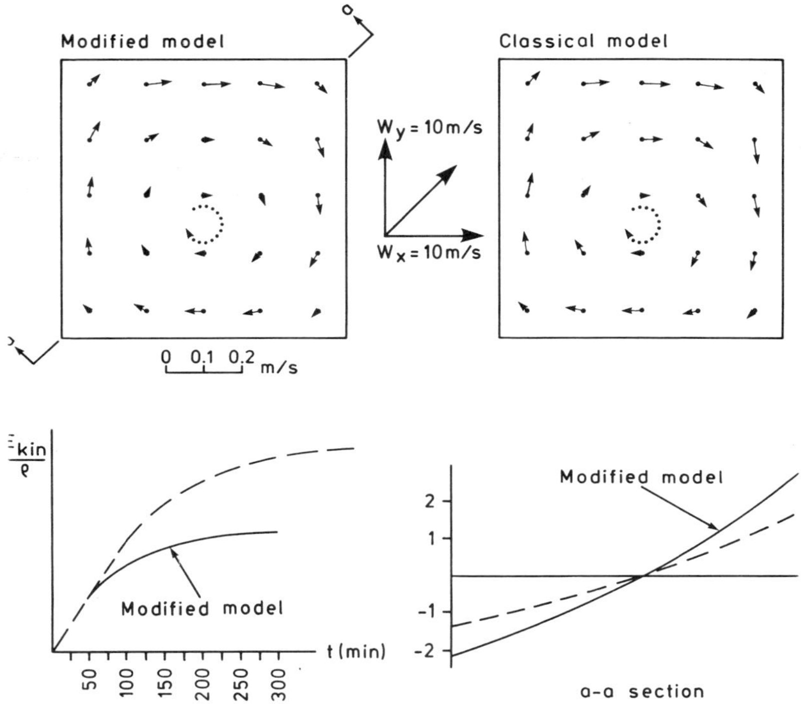

Fig. 2.16 Circulation patterns, energy evolution and surface profile for the classical and modified wind generated circulation model

shoal, diffract or break. This circulation is due to the spatial variation in the momentum contained by the waves. As was mentioned in Chapter 1, the radiation stresses describe the depth mean wave momentum integrated over a wave period T. It was shown in Section 1.7 (Equations (1.76)–(1.78)) that the radiation stress components $\sigma_{xx}, \sigma_{xy}, \sigma_{yx}, \sigma_{yy}$ form a symmetric second order tensor and their action is completely analogous to the stress tensor. As they describe the depth mean components of momentum along the sides of an infinitesimal column of water (base dx, dy, height h), their inclusion in the mathematical model for the 2DH flows is straightforward. Their spatial gradients appear on the right hand side of the equilibrium equations. The wave generated circulation model containing the radiation stresses as flow forcing factor has the form:

$$\frac{\mathrm{d}U}{\mathrm{d}t} = -g\frac{\partial\zeta}{\partial x} - \frac{\tau_{bx}}{\rho h} - \frac{1}{\rho h}\left(\frac{\partial\sigma_{xx}}{\partial x} + \frac{\partial\sigma_{xy}}{\partial y}\right) \tag{2.73}$$

$$\frac{\mathrm{d}V}{\mathrm{d}t} = -g\frac{\partial \zeta}{\partial y} - \frac{\tau_{by}}{\rho h} - \frac{1}{\rho h}\left(\frac{\partial \sigma_{xy}}{\partial x} + \frac{\partial \sigma_{yy}}{\partial y}\right) \tag{2.74}$$

$$\frac{\partial \zeta}{\partial t} + \frac{\partial (Uh)}{\partial x} + \frac{\partial (Vh)}{\partial y} = 0 \tag{2.75}$$

The $\zeta(x, y, t)$ function describes the difference between the mean surface level (with respect to waves) and the still water level (difference between MWL and SWL). Its time variation does not refer to the rapid variation of the surface during the passage of the waves (with period of some seconds) but to the long term variations of the MWL. The Coriolis term can be dropped from (2.73), and (2.74) when the model is applied over small geophysical domains such as narrow coastal strips where the refraction or diffraction of waves takes place.

A criticism that can be made of the form of the model presented refers to the depth averaging of the momentum. Unlike long-wave generated barotropic flows and wind generated circulation, the velocity variation over the depth is considerable (as analysis in the x, y, z, t space and measurements in the surf zone indicate) and the depth varying model should be used. It is beyond the scope of this book and most of the operational models used today. The bed shear is expressed also by quadratic relations of the form (2.59).

The remaining boundary conditions and the numerical solution follow the same path as the described wind or tidal circulation. The difference appears in the statements for the composition of UN, VN values, where the radiation stress gradients appear. This means that a subprogram for the computation of the SXX(I, J), SYY(I, J) and SXY(I, J) components from the wave characteristics should precede the computation of velocities. First, the value of the wave parameter n, the wave angle θ (angle measured from the positive Ox axis to the wave orthogonal according to the notation of Fig. 2.17) and the wave height H (the H_{rms} value in case of random waves) have to be computed. The radiation stress components computed refer to the mesh centers.

Program 11 is adapted to a flow domain of variable depth bounded by an upper coastal boundary, a lower open sea boundary and lateral uniform flow boundaries ($\partial/\partial x = 0$). It may also contain several coastal structures, such as moles, a breakwater, etc. around which the wave generated circulation has to be computed. A typical morphology of the flow domain is given in Fig. 2.17.

The order of magnitude of the Δx step in such a situation (circulation around a coastal structure) is smaller than that for a model of general circulation in a bay $O[\Delta x] = 10^2$ m. It is preferable

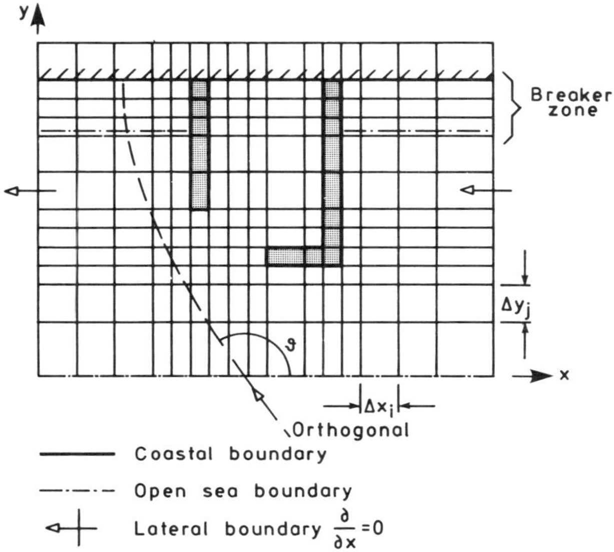

Fig. 2.17 Wave induced circulation in refraction-diffraction and breaking areas. Morphology of domain

for the Δx, Δy steps to be variable, so that the resolution is higher in desired areas of the flow domain.

PROGRAM 11: 2-D WIND AND WAVE GENERATED CIRCULATION MODEL

```
5REM 2-D WIND AND WAVE GENERATED CIRCULATION M
ODEL
10DIMU(20,20),UN(20,20),V(20,20),VN(22,20),Z(20
,20),H(20,20),XX(20,20),XY(20,20),YY(20,20),DX(20)
,DY(20),IB(80),JB(80),NB(80)
20READDT,IM,JM,NM,CF,WX,WY,CS,TH,KB
22DATA...
30FORI=1TOIM-1:READDX(I):NEXTI
40DATA...
50FORJ=1TOJM-1:READDY(J):NEXTJ
60DATA...
70FORI=1TOIM-1:FORJ=1TOJM-1:READH(I,J),XX(I,J),
XY(I,J),YY(I,J) :NEXTJ:NEXTI
75DATA...
80SX=CS*WX*SQR(WX^2+WY^2):SY=CS*WY*SQR(WX^2+WY^
2)
90FORK=1TOKB:READIB(K),JB(K),NB(K):NEXTK
100DATA....
110FORI=1TOIM-1:H(I,0)=H(I,1):XX(I,0)=XX(I,1):YY
```

```
(I,0)=YY(I,1):XY(I,0)=XY(I,1):DY(0)=DY(1):NEXTI
  120FORJ=1TOJM-1:H(0,J)=H(1,J):XX(0,J)=XX(1,J):XY
(0,J)=XY(1,J):YY(0,J)=YY(1,J):DX(0)=DX(1):H(IM,J)=
H(IM-1,J):XX(IM,J)=XX(IM-1,J):YY(IM,J)=YY(IM-1,J):
XY(IM,J)=XY(IM-1,J):DX(IM)=DX(IM-1):NEXTJ
  130FORK=1TOKB:IF NB(K)<>4 THEN GOTO150
  140I=IB(K):J=JB(K):H(I,J)=1
  150NEXTK:N=0:T=0
  160N=N+1:T=T+DT
  170FORJ=1TOJM-2:FORI=1TOIM-1:Z(I,J)=Z(I,J)-DT*((
U(I+1,J)*(H(I+1,J)+H(I,J))-U(I,J)*(H(I,J)+H(I-1,J)
))/DX(I)+(V(I,J+1)*(H(I,J+1)+H(I,J))-V(I,J)*(H(I,J
)+H(I,J-1)))/DY(J))/2 :NEXTI:NEXTJ
  180FORJ=1TOJM-2:FORI=2TOIM-1:HM=(H(I,J)+H(I-1,J)
)/2:VV=(V(I,J)+V(I,J+1)+V(I-1,J)+V(I-1,J+1))/4:XM=
(DX(I)+DX(I-1))/2:YM=(DY(J)+(DY(J-1)+DY(J+1))*.5)

  190UN(I,J)=U(I,J)*TH+(1-TH)/4*(U(I+1,J)+U(I-1,J)
+U(I,J+1)+U(I,J-1))-DT*(U(I,J)*(U(I+1,J)-U(I-1,J))
/XM/2+VV*(U(I,J+1)-U(I,J-1))/YM+9.81*(Z(I,J)-Z(I-1
,J))/XM+((XX(I,J)-XX(I-1,J))/XM+(XY(I,J+1)+XY(I-1,
J+1)-XY(I,J-1)-XY(I-1,J-1))/YM/2)/HM)
  200UN(I,J)=UN(I,J)-DT*CF*U(I,J)*SQR(U(I,J)^2+VV^
2)/HM+DT*SX/HM:NEXTI:NEXTJ
  210FORJ=2TOJM-2:FORI=1TOIM-1:UU=(U(I,J)+U(I+1,J)
+U(I,J-1)+U(I+1,J-1))/4:HM=(H(I,J)+H(I,J-1))/2:XM=
DX(I)+(DX(I+1)+DX(I-1))*.5:YM=(DY(J)+DY(J-1))/2
  220VN(I,J)=V(I,J)*TH+(1-TH)/4*(V(I,J+1)+V(I,J-1)
+V(I+1,J)+V(I-1,J))-DT*(V(I,J)*(V(I,J+1)-V(I,J-1))
/YM/2+UU*(V(I+1,J)-V(I-1,J))/XM+9.81*(Z(I,J)-Z(I,J
-1))/YM+((YY(I,J+1)-YY(I,J-1))/YM+(XY(I+1,J)+XY(I+
1,J-1)-XY(I-1,J)-XY(I-1,J-1))/2/XM)/HM)
  230VN(I,J)=VN(I,J)-DT*CF*V(I,J)*SQR(V(I,J)^2+UU^
2)/HM+DT*SY/HM:NEXTI:NEXTJ
  240FORK=1TOKB:I=IB(K):J=JB(K)   :ONNB(K)-1 GOTO25
0,260,270
  250UN(I,J)=0:GOTO280
  260VN(I,J)=0:GOTO280
  270VN(I,J)=0:UN(I,J)=0
  280NEXTK
  290FORJ=1TOJM-1:UN(1,J)=UN(2,J):VN(0,J)=VN(1,J):
UN(IM,J)=UN(IM-1,J):VN(IM,J)=VN(IM-1,J):NEXTJ
  300FORI=1TOIM-1:UN(I,0)=UN(I,1):VN(I,1)=-Z(I,1)*
SQR(9.8/H(I,1)):NEXTI
  310FORJ=1TOJM:FORI=1TOIM:U(I,J)=UN(I,J):V(I,J)=V
N(I,J):NEXTI:NEXTJ
  320IF N/20<>INT(N/20) THEN GOTO160
  330EKK=EK:EK=0:FORJ=1TOJM-2:FORI=1TOIM-1:EK=EK+(
(U(I,J)+U(I+1,J))^2+(V(I,J)+V(I,J+1))^2)/8/DX(I)*D
Y(J)*H(I,J)  :NEXTI:NEXTJ
```

```
  340PRINTN,EK
  350IF N<NM OR ABS(EKK-EK)/EK>.001 THEN GOTO 160
  360FORJ=JM-2 TO 2 STEP-1:FORI=1TOIM-1:PRINT(U(I,
J)+U(I+1,J))/2;:NEXTI:PRINT:NEXTJ:PRINT
  370FORJ=JM-2 TO 2 STEP-1:FORI=1TOIM-1:PRINT(V(I,
J)+V(I,J+1))/2;:NEXTI:PRINT:NEXTJ:PRINT
  380FORJ=JM-2 TO 2 STEP-1:FORI=1TOIM-1:PRINTZ(I,J
);:NEXTI:PRINT:NEXTJ:PRINT
  400END
```

In the description of variables the only new ones, compared with the previous models, are the data arrays for the radiation stresses $\sigma_{xx}/\rho = \mathrm{XX}$, $\sigma_{xy}/\rho = \mathrm{XY}$, $\sigma_{yy}/\rho = \mathrm{YY}$ and the parameter θ in the recommended diffusive Lax type finite difference for the time derivative.

$$\frac{\partial U}{\partial t} = [U_{ij}^{n+1} - U_{ij}^{n} \cdot \theta + \{(1-\theta)/4\} \times (U_{i+1j}^{n} + U_{i-1j}^{n} + U_{ij+1}^{n} + U_{ij-1}^{n})]/\Delta t \quad (2.76)$$

This type of finite difference approximation entails horizontal momentum diffusion with a diffusion coefficient equal to $\{(1-\theta)\Delta x^2\}/\Delta t/4$. Care should be taken in the determination of θ that the introduced diffusion is realistic.

The application is concerned with the wave generated circulation in a coastal area of simple geometry (i.e. straight coastline, constant bed slope and parallel bed contours). An L-shaped mole is constructed to face the prevailing waves. The flow domain and its discretisation are shown in Fig. 2.18. The waves approaching the coast have $H_0 = 2$ m, $T = 7$ s and direction SW.

The main program data are: DT = 5 s, IM = 20, VM = 16, MN = 300, CF = 0.01, TH = 0.95, KB = 26.

DX(I): 40, 40, 40, 40, 20, 10, 10, 10, 10, 10, 20, 20, 20, 20, 20, 40, 40, 40, 40, 40

DY(J): 40, 40, 20, 20, 20, 10, 10, 10, 10, 10, 10, 10, 10, 10, 10

The water depth refers to the mesh centers. In the modelled port basin the depth is considered constant, equal to 6 m. The radiation stresses are computed via the refraction and diffraction of waves in the area. The B's (orthogonal spacing) giving k_r and the θ's (wave angles) are computed in the flow domain. In the basin interior k_D and θ are computed by the Wiegel tables (diffraction around an infinite breakwater). The above refraction and diffraction computations give the wave heights and θ at the mesh centers. The σ_{xx}/ρ, σ_{xy}/ρ, σ_{yy}/ρ values are computed by means of a sub-routine for radiation stress.

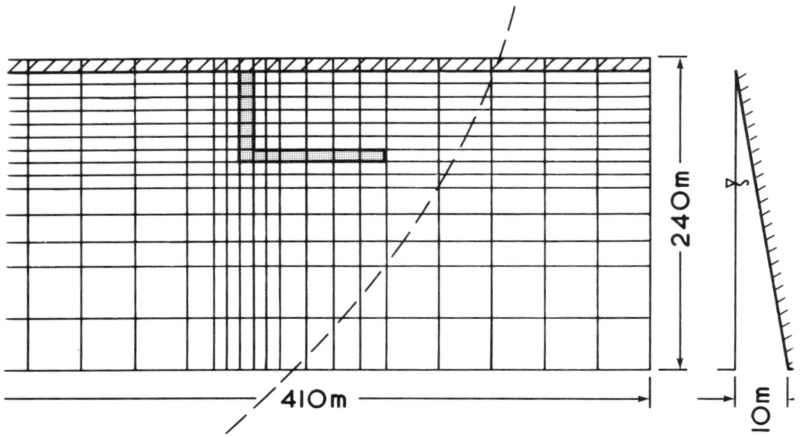

Fig. 2.18 Flow domain, its discretisation and typical refracted orthogonal

PROGRAM 12: RADIATION STRESS COMPUTATION FROM WAVE ANGLE AND ORTHOGONALS DISTANCE

```
    5REM RADIATION STRESSES COMPUTATION FROM WAVE
ANGLE AND ORTHOGONALS DISTANCE
   10DIMH(20,20),HH(20,20),SXX(20,20),SXY(20,20),S
YY(20,20),TH(20,20),DX(20),DY(20),B(20,20),LL(50)
   20READIM,JM,HO,T,DM,BO
   30DATA...
   40FORJ=1TOJM-1:FORI=1TOIM-1:READH(I,J):NEXTI:NE
XTJ
   50DATA...
   60FORJ=1TOJM-1:FORI=1TOIM-1:READTH(I,J):NEXTI:N
EXTJ
   70DATA...
   80FORJ=1TOJM-1:FORI=1TOIM-1:READB(I,J):NEXTI  :N
EXTJ
   90DATA...
  100LO=9.81*T^2/2/PI
  105FORK=1TODM:L=LO:TM=L
  110MN=(L+TM)/2:TM=L:A=2*PI*K/MN:L=LO*(EXP(A)-EXP
(-A))/(EXP(A)+EXP(-A))
  120IF(ABS(L-TM)>.1)THEN GOTO110
  130LL(K)=L:NEXTK
  140FORJ=1TOJM:FORI=1TOIM  :IF(H(I,J)=0)THEN GOTO1
95
  150FORK=1TODM  :IF(H(I,J)<K)THEN GOTO170
  160NEXTK
  170L=LL(K-1)+(LL(K)-LL(K-1))*(K-H(I,J)):A=2*2*PI
```

```
*H(I,J)/L:SANH=(EXP(A)-EXP(-A))/2:N=.5*(1+A/SANH):
KS=SQR(LO/2/N/L):KR=SQR(BO/B(I,J)):HH(I,J)=HO*KS*K
R
  180IF(HH(I,J)>.7*H(I,J))THEN HH(I,J)=.7*H(I,J)
  190SXX(I,J)=9.8*HH(I,J)^2/16*(2*N-1+N*(COS(RAD(T
H(I,J))))^2):SXY(I,J)=9.81*HH(I,J)^2/16*N*SIN(2*RA
D(TH(I,J))):SYY(I,J)=9.81*HH(I,J)^2/16*(2*N-1+N*(S
IN(RAD(TH(I,J))))^2)
  195NEXTI:NEXTJ
  200FORJ=1TOJM-1:FORI=1TOIM-1:PRINTH(I,J),HH(I,J)
,SXX(I,J),SXY(I,J),SYY(I,J):NEXTI:NEXTJ
  210END
```

Description of variables:

IM, JM = maximum values of indices along Ox, Oy
H0 = wave height in the open sea
T = wave period
DM = maximum water depth in the flow area in m
B0 = wave orthogonals distance in the open sea
H(I, J) = water depth values in mesh center
TH(I, J) = corresponding wave approach angle measured counter-clockwise from the x^+ axis in degrees
B(I, J) = corresponding wave orthogonals distance.

In this routine first the wave characteristics L, H, n are computed at the mesh centers and then the radiation stress components. Near the coast, the wave breaking is checked and the wave height is estimated with a breaking factor $\gamma = 0.7$ ($H = \gamma h$). The radiation stress components are introduced as data to the circulation program. The current is computed up to the stabilisation of the flow. The steady current pattern is illustrated in Fig. 2.19. Some easily discernible eddies are shown. The strong longshore current is present in the surf zone before the mole and is re-established some distance after it.

The wave generated current field diminishes rapidly away from the coast. This type of circulation, as stated a priori, is important in limited areas where the water depth is much smaller than the wavelength and intensive wave deformation is taking place. Nevertheless, it is an important feature in the design and operation of coastal structures.

2.5 DENSITY CURRENTS. STRATIFIED FLOWS

2.5.1 General notions, definitions

In all the previous circulation models the water was assumed

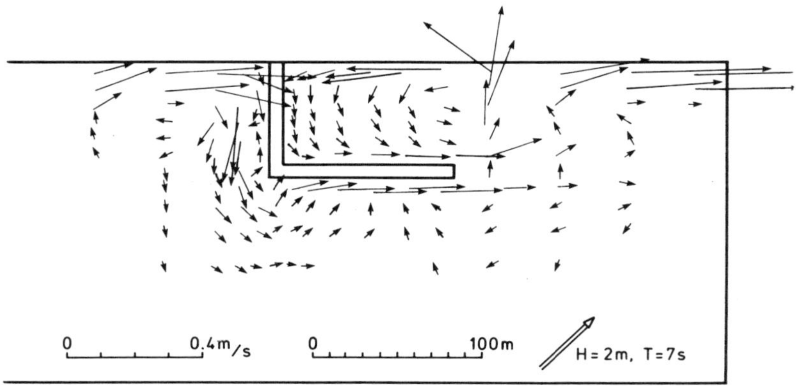

Fig. 2.19 Wave-induced circulation around the mole

homogeneous (of constant density) and the gravity force distributed uniformly throughout the mass of the fluid. The resulting hydrostatic pressure distribution in the case of quasi-horizontal flows was assumed linear and the horizontal pressure gradient was expressed by means of the free surface elevation gradient

$$-\frac{1}{\rho}\cdot\frac{\partial p}{\partial x} = -g\frac{\partial \zeta}{\partial x} \tag{2.77}$$

In the case of nonhomogeneous fluid, however, the density varies in three-dimensional space $\rho(x, y, z, t)$. The variation is due to:

(1) salinity variation, as in the case where fresh land-based water is mixed with sea water, and
(2) horizontal or vertical temperature differences.

The resulting density field ($\rho = \rho(S, T)$) implies a hydrostatic pressure distribution of a more complicated form:

$$p(z) = \int_z^{\zeta} \rho g \, \mathrm{d}z + p_0 \tag{2.78}$$

where p_0 is the atmospheric pressure. The general form of the horizontal pressure gradient becomes

$$-\frac{1}{\rho}\cdot\frac{\partial p}{\partial x} = -\frac{g}{\rho}\int_z^{\zeta}\frac{\partial \rho}{\partial x}\mathrm{d}z - \frac{g}{\rho}\cdot\frac{\partial \zeta}{\partial x}\rho\bigg|_{z=\zeta} - \frac{1}{\rho}\cdot\frac{\partial p_{\mathrm{atm}}}{\partial x} \tag{2.79}$$

In the case where depth averaging provides a realistic approximation (intensive vertical mixing due to turbulence) and

$\rho = \rho(x, y, t)$ then (2.79) becomes

$$-\frac{1}{\rho}\cdot\frac{\partial p}{\partial x} = -\frac{g}{\rho}\cdot\frac{\partial \rho}{\partial x}(\zeta - z) - \frac{1}{\rho}\cdot\frac{\partial p_{\text{atm}}}{\partial x} - g\frac{\partial \zeta}{\partial x} \tag{2.80}$$

When $\rho =$ constant we return to the form of Equation (2.77).

In a nonhomogeneous fluid either the nonhomogeneity is itself the flow generating factor (density currents) or it influences the hydrodynamic conditions in the flow domain. Two limiting cases can be distinguished in nonhomogeneous flow domains:

(1) Well-mixed domains in which water is homogeneous over the depth. Horizontal density gradients may exist resulting in horizontal flow, a form of density currents.

(2) Fully stratified domains. Two or more distinct layers are formed along the vertical, separated by a thin interface (pycnocline). The layers communicate only through their interface where mixing phenomena are considered weak. In a two-layers environment when the more dense layer is above the less dense, the case is hydrodynamically unstable and vertical convection may take place resulting in mixing of the two layers.

When the less dense layer is on top, the situation is hydrodynamically stable and a relatively large energy supply is required to achieve the mixing of the two layers (storm waves for example).

When the surface and the interface are horizontal the two layers may be stagnant. Horizontal gradients in the elevation of the two surfaces (free and interface) result in fluid motion known as baroclinic motion.

A mathematical model of coastal circulation in the general form consisting of Equations (2.5) and (2.6) can describe the circulation of a nonhomogeneous fluid. The differences concern the horizontal pressure gradients and are expressed by Equation (2.79). This means that the density magnitude must be known over the whole extent of the flow domain. The density may be considered variable in space and constant in time during the development of the circulation phenomenon or may be considered as an evolving magnitude. In the later case, as the density is basically a function of temperature T and salinity S, two equations formulated on the principle of heat and salt conservation are required to complete the model; their form is the same. For $c = T$ or $c = S$, the conservation equation is

$$\frac{\partial c}{\partial t} + \frac{\partial (cu)}{\partial x} + \frac{\partial (cv)}{\partial y} + \frac{\partial (cw)}{\partial z} = \frac{\partial}{\partial x}\left(D_h\frac{\partial c}{\partial x}\right) + \frac{\partial}{\partial y}\left(D_h\frac{\partial c}{\partial y}\right) + \frac{\partial}{\partial z}\left(D_v\frac{\partial c}{\partial z}\right) \tag{2.81}$$

Equation (2.81) is the advective turbulent diffusion equation for a conservative substance with different horizontal and vertical diffusion coefficients D_h, D_v. The state equation relating ρ to S and T can be approximated by a linear one

$$\rho = \rho_0(1 + ac) \tag{2.82}$$

where a in the case of the $\rho(T)$ function is a volume expansion coefficient. A final difference between the homogeneous and nonhomogeneous fluid circulation model concerns the eddy viscosity magnitude distribution. This magnitude describes the rate of diffusion of momentum over the depth. In the case of a steep density gradient along the interface the momentum cannot be diffused through the pycnocline. The parametrisation of that physical process, i.e. the tapering of momentum in one layer is done through the eddy viscosity distribution. A nondimensional parameter called the Richardson number is the parameter involved, defined as the ratio

$$R_i = \frac{g\dfrac{\partial \rho}{\partial z}}{\rho\left(\dfrac{\partial u}{\partial z}\right)^2} \tag{2.83}$$

The large density gradients on the pycnocline results in high R_i. The quantification of the influence of R_i on the eddy viscosity is done usually by means of the relation

$$v_v = v_{v_{\text{homog}}} \cdot (1 + aR_i)^{-b} \tag{2.84}$$

where $v_{v_{\text{homog}}}$ is the eddy viscosity for the homogeneous fluid and a and b coefficients fixed by in situ measurements, $O[a] = 10$, $O[b] = 0.5$.

The density distribution, eddy viscosity and velocity in the case of steady uniform barotropic flow in a stratified fluid is shown schematically in Fig. 2.20. It is obvious that the two flow regions (the

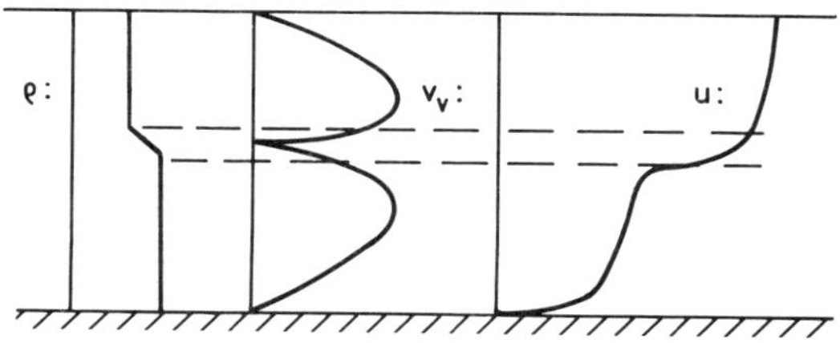

Fig. 2.20 Eddy viscosity and velocity distribution for barotropic llow in stratified medium

upper and lower) behave differently under the same flow generating factor. The momentum diffusion is minimised across the interface and as a result sharp velocity gradients develop there. The flow profile in each layer is almost uniformly distributed over the depth of the respective layer, but the velocity in the upper layer is higher than that in the lower layer due to the greater friction at the bed than at the interface.

From the above observations it becomes clear that the first simplifying approximation for the description of the phenomenon of a 2-layered flow is to assume two distinct layers and apply the depth averaged model for each one separately.

As the solution of the depth-varying model for such types of barotropic–baroclinic flow is beyond the scope of this book we will limit our modelling to the 2DH methodology applied to the two limiting cases of horizontal and vertical density variability. First, the case of fluid fully mixed over the depth, with horizontal density variation; second, the case of vertically layered fluid (2-layer model).

2.5.2 A model for horizontal flow of a nonhomogeneous fluid vertically well mixed

This model is the crudest approximation for the description of circulation with density variations due to advection and diffusion in a coastal domain of very shallow water connected to the open sea (an infinite domain of constant density) and receiving fresh water outflow from a land based source. It is assumed that due to several physical reasons vertical mixing is complete and that only horizontal density differences drive or regulate the circulation.

Let us confine our investigation for simplicity to the one-dimensional case. The general form of the mathematical model for the depth varying circulation is written in the unknowns $u(x, z, t)$, $\zeta(x, t)$, $\rho(x, t)$, $S(x, t)$

$$\frac{\partial u}{\partial t} + u\frac{\partial u}{\partial x} = -g\frac{\partial \zeta}{\partial x} + g\frac{z}{\rho}\frac{\partial \rho}{\partial x} + \frac{\partial}{\partial z}\left(v\frac{\partial u}{\partial z}\right) \tag{2.85}$$

$$\frac{\partial \zeta}{\partial t} + \frac{\partial (Uh)}{\partial x} = q \qquad \left(U = \frac{1}{h}\int_{-h}^{0} u\,\mathrm{d}z\right) \tag{2.86}$$

where q is the inflowing discharge of fresh water of density ρ_0,

$$\rho = \rho_0(1 + aS) \tag{2.87}$$

$$\frac{\partial S}{\partial t} + \frac{\partial (SU)}{\partial x} = \frac{\partial}{\partial x}\left(R\frac{\partial S}{\partial x}\right) \tag{2.88}$$

The boundary conditions completing (2.85)–(2.87) are $u|_{z=-h}=0$, $\partial u/\partial z|_{z=0}=0$ with respect to velocity and $\partial S/\partial x=0$ on no-flow coastal boundaries, $S=$ given constant on an open sea or river boundary with respect to salinity. The parameters v_v and R (vertical eddy viscosity and horizontal mass dispersivity) are also unknown magnitudes. A sensitivity analysis shows that their values influence the solutions substantially. An order of magnitude analysis for them indicate that

$$O[v_v]=O[U]\cdot O[h]\cdot 0.1 \tag{2.89}$$

$$O[R]=O[U]\cdot O[S]\cdot O[\Delta x/\Delta s] \tag{2.90}$$

where $\Delta s/\Delta x$ a measure of the horizontal salinity gradient.

A substantial simplification is introduced by integration of Equation (2.85) over the depth and its expression in terms of the depth mean velocity

$$\frac{\partial U}{\partial t}+U\frac{\partial U}{\partial x}=-g\frac{\partial \zeta}{\partial x}-\frac{gh}{2}\cdot\frac{\partial \ln \rho}{\partial x}-\frac{kU}{h} \tag{2.91}$$

The bed friction term can have this linear form or a classical quadratic one.

The model consisting of Equations (2.91), (2.86), (2.88) and the state Equation (2.87) is capable of describing the formation of density induced currents and the density field and their continuous interaction in the case of horizontal density variations only. These density currents can be induced by the inflow of fresh water into initially stagnant salt water or by the joining of a fresh water body with the sea.

The problem of the difference in the order of magnitude of the time scales in the formation of the hydrodynamic and density magnitudes can be resolved by the use of implicit finite differences for (2.91) permitting the increase in a common Δt step used for both (2.91) and (2.88).

The numerical solution of the depth averaged horizontally stratified fluid model presented here is based on the explicit finite difference scheme using the staggered grid illustrated in Fig. 2.21. The velocity values are computed at the cross-sections (nodes) while the water depth, surface elevation and salinity-density values refer to the reaches between successive nodes. The finite difference approximation of the model's equations are:

(1) Continuity equation:

$$\frac{\zeta_i^{n+1}-\zeta_i^n}{\Delta t}=\frac{1}{2\Delta x}((h_{i+1}+h_i)U_{i+1}^n-(h_i+h_{i-1})U_i^n) \tag{2.92}$$

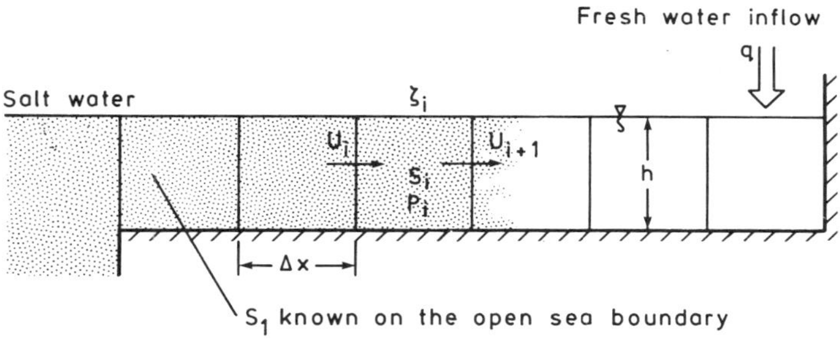

Fig. 2.21 1-D density current due to horizontal density gradients. Domain morphology and discretisation

(2) Equilibrium equation:

$$\frac{U_i^{n+1}-U_i^n}{\Delta t} = -\frac{U_i^n}{2\Delta x}(U_{i+1}^n - U_{i-1}^n) - \frac{g}{\Delta x}(\zeta_i^n - \zeta_{i-1}^n) - \frac{g(h_i+h_{i-1})}{4\Delta x}$$
$$\times (\ln \rho_i^n - \ln \rho_{i-1}^n) - 2kU_i^n/(h_i+h_{i-1}) \quad (2.93)$$

(3) Salt conservation:

$$\frac{S_i^{n+1}-S_i^n}{\Delta t} = -\frac{1}{g\Delta x}[(S_{i+1}^n + S_i^n)U_{i+1}^n - (S_i^n + S_{i-1}^n)U_i^n]$$
$$+\frac{R}{\Delta x^2}(S_{i+1}^n - 2S_i^n + S_{i-1}^n) \quad (2.94)$$

(4) State equation:

$$\rho_i^{n+1} = \rho_0(1+\alpha S_i^{n+1}) \quad (2.95)$$

The procedure is organised in a BASIC program (Program 13) referring to the case of an enclosed domain (upstream and downstream no-flow conditions $\partial S/\partial x = 0$, $U = 0$) where an initial density distribution induces circulation that, after the completion of mixing, leads asymptotically to a new equilibrium after the homogenisation of the fluid and the frictional decay of the initial motion. Different boundary conditions can be used with a slight modification to the program to simulate other types of density induced flow.

PROGRAM 13: 1-D CURRENT WITH HORIZONTAL DENSITY GRADIENT

```
    5REM 1-D DENSITY CURRENT HORIZONTAL DENSITY GR
ADIENT
   10DIMU(21),UN(21),Z(21),S(21),SN(21),R(21),H(21
)
```

```
20READ DT,DX,IM,NM,K,E,H0,A
30DATA...
40FORI=0TOIM:READS(I):R(I)=100*(1+A*S(I)):NEXTI
42DATA...
45FORI=0TOIM:READH(I):NEXTI
47DATA...
50N=0:T=0
60N=N+1:T=T+DT
70FORI=1TOIM-1:Z(I)=Z(I)-DT/2/DX*((H(I)+H(I+1))
*U(I+1)-(H(I)+H(I-1))*U(I)):NEXTI
80FORI=2TOIM-1:UN(I)=U(I)-DT/2/DX*U(I)*(U(I+1)-
U(I-1))-DT*9.81/DX*(Z(I)-Z(I-1))-DT*9.81*(H(I)+H(I
-1))/4*(LOG(R(I))-LOG(R(I-1)))/DX-DT*K*U(I)*ABS(U(
I))*2/(H(I)+H(I-1)):NEXTI
90FORI=1TOIM-1:SN(I)=S(I)-DT/2/DX*((S(I+1)+S(I)
)*U(I+1)-(S(I)+S(I-1))*U(I))+DT/DX^2*E*(S(I+1)-2*S
(I)+S(I-1)):NEXTI:SN(0)=SN(1):SN(IM)=SN(IM-1)
100FORI=0TOIM:U(I)=UN(I):S(I)=SN(I):R(I)=100*(1+
A*S(I)):NEXTI
110IFN/20<>INT(N/20) THEN GOTO 60
120PRINTN
130FORI=1TOIM:PRINTZ(I);:NEXTI:PRINT
140FORI=1TOIM:PRINTU(I);:NEXTI:PRINT
150FORI=1TOIM:PRINTS(I);:NEXTI:PRINT
160IFN<NM THEN GOTO 60
170END
```

Description of main variables:

DT, DX	= time and space discretisation steps
IM, NM	= number of cross-sections in the flow domain discretisation and maximum number of time steps
K	= bed friction coefficient (m/s)
E	= salt dispersion coefficient (m^2/s)
H0	= common water depth (in this case $h = cte$, variable in general)
A	= linear $\rho - S$ relation coefficient
S0	= salinity at (seaward) boundary (kg/m^3).

The application refers to a flow domain of constant depth H0 = 3.67 m. length 20 km, bounded by coastal boundaries where the density has an initial linear variation; on the left boundary the water is fresh, $\rho = 1000\ kg/m^3$, and on the right it is salt sea water, $\rho = 1020\ kg/m^3$. The horizontal density variation induces density currents and free surface variations that result in the seiching of the flow domain at the lowest natural frequency. The phenomenon proceeds up to the homogenisation of the fluid after $\approx 10{,}000$ s.

The application data are: DT = 10 s, DX = 100 m, IM = 21, NM = 1000, K = 0.001 m/s, E = 100 m^2/s, H0 = 3.67 m, A = 0.0007, S0 = 30 ppt. The flow domain, its discretisation, the velocity evolution in the middle $U_B(t)$ and the $\zeta_A(t)$ evolution at the boundary are depicted in Fig. 2.22(a). Figure 2.22(b) shows the change in salinity along the flow domain up to homogenisation.

2.5.3 Stratified flow model (2-layer model)

In the case of vertically stratified fluid the model describing the depth averaged magnitudes is not realistic. Either the depth-varying model describing the continuous evolution of the hydrodynamic magnitudes $u(x, y, z, t)$ along the depth, or a series of superimposed depth-averaged models corresponding to each layer have to be used. The second approach is common operational practice and its 2-layer version will be presented here.

In the case of the existence of a discrete pycnocline, the fluid can be considered as consisting of two quasi-horizontal layers and, according to the notation of Fig. 2.23, the equations of equilibrium of forces and mass continuity can be written for the two layers separately. For the pressure in the upper layer,

$$p = \lambda \rho_0 g(\zeta - z) \rightarrow \frac{1}{\lambda \rho_0} \cdot \frac{\partial p}{\partial x} = g \frac{\partial \zeta}{\partial x} \tag{2.96}$$

while for the lower layer it can be easily proved that

$$p_0 = \lambda \rho_0 g(\zeta + h - \zeta_0) + \rho_0 g(-z - h + \zeta_0) \tag{2.97}$$

giving

$$\frac{1}{\rho_0} \cdot \frac{\partial p_0}{\partial x} = \lambda g \frac{\partial \zeta}{\partial x} - \lambda g \frac{\partial \zeta_0}{\partial x} + g \frac{\partial \zeta_0}{\partial x}$$

$$= \lambda g \frac{\partial \zeta}{\partial x} + (1 - \lambda) g \frac{\partial \zeta_0}{\partial x} \tag{2.98}$$

where λ is the ratio of the densities of the two layers:

$$\lambda = \frac{\rho}{\rho_0} < 1 \qquad (1 \geqslant \lambda \geqslant 0.98) \tag{2.99}$$

From Equation (2.98) it is obvious that the free surface gradients multiplied by λ ($\simeq 1$) and the interface gradients multiplied by $(1 - \lambda)$ ($\simeq 0$), are involved in the determination of the velocity field in the lower layer. This indicates a priori that the orders of magnitude of the ζ and ζ_0 values have to be different so that the two pressure terms keep the same order, i.e. $O[\zeta] < O[\zeta_0]$.

A second important concern in a stratified fluid model is the

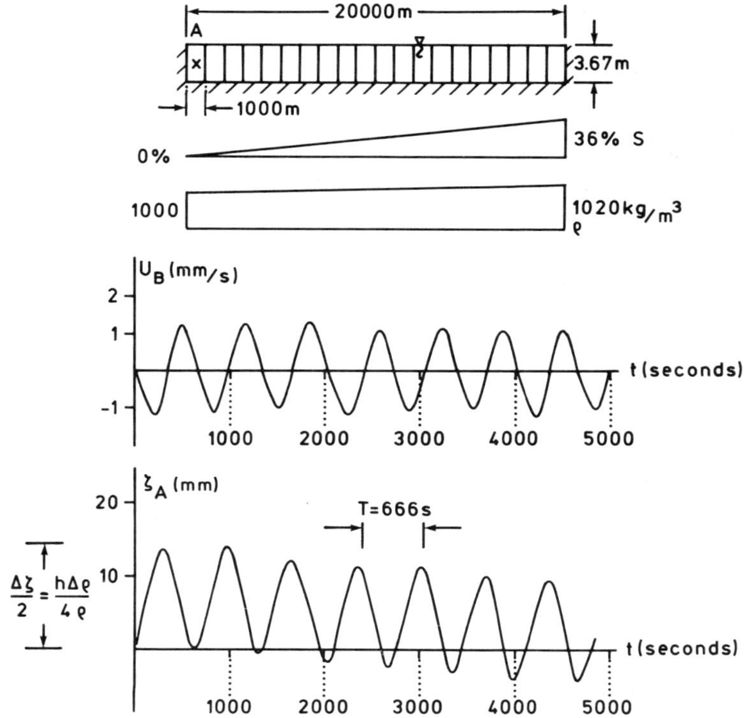

Fig. 2.22(a) *1-D density current flow domain, its discretisation and evolution of velocity and surface elevation at two locations*

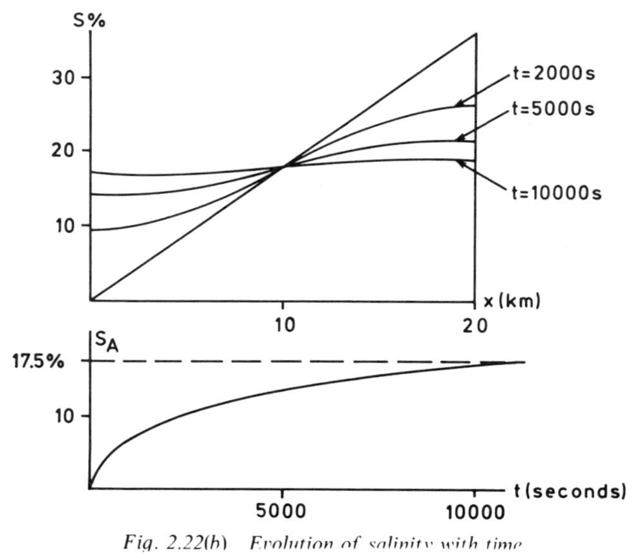

Fig. 2.22(b) *Evolution of salinity with time*

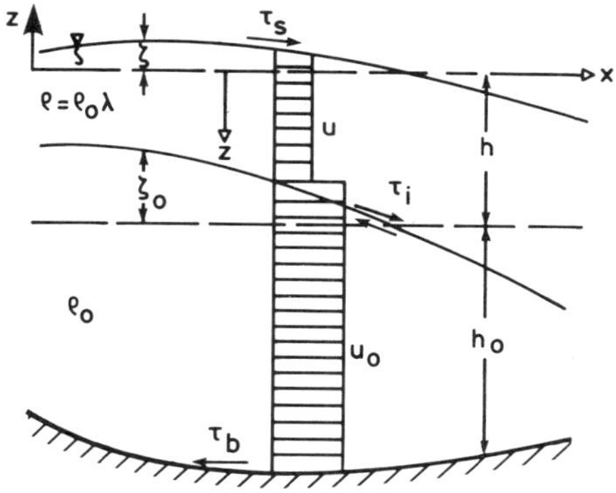

Fig. 2.23 Stratified flow model. Basic notations

development of interfacial stresses. Their magnitude can be simply approximated by a quadratic form containing the difference of the mean velocities of the two layers. The form of the 2DH, 2-layer, model is

(1) Continuity equations in the two layers:

$$\frac{\partial \zeta_0}{\partial t}+\frac{\partial}{\partial x}(U_0 h_0)+\frac{\partial}{\partial y}(V_0 h_0)=0 \tag{2.100}$$

$$\frac{\partial(\zeta-\zeta_0)}{\partial t}+\frac{\partial}{\partial x}(Uh)+\frac{\partial}{\partial y}(Vh)=0 \tag{2.101}$$

(2) Equilibrium of forces equations along x, y in the two layers:

$$\frac{\partial U}{\partial t}+U\frac{\partial U}{\partial x}+V\frac{\partial U}{\partial y}=-g\frac{\partial \zeta}{\partial x}+\frac{\tau_{sx}-\tau_{ix}}{\lambda\rho_0 h}+fV \tag{2.102}$$

$$\frac{\partial V}{\partial t}+U\frac{\partial V}{\partial x}+V\frac{\partial V}{\partial y}=-g\frac{\partial \zeta}{\partial y}+\frac{\tau_{sy}-\tau_{iy}}{\lambda\rho_0 h}-fU \tag{2.103}$$

$$\frac{\partial U_0}{\partial t}+U_0\frac{\partial U_0}{\partial x}+V_0\frac{\partial U_0}{\partial y}=-\lambda g\frac{\partial \zeta}{\partial x}-(1-\lambda)g\frac{\partial \zeta_0}{\partial x}+\frac{\tau_{ix}-\tau_{bx}}{\rho_0 h_0}+fV_0 \tag{2.104}$$

$$\frac{\partial V_0}{\partial t}+U_0\frac{\partial V_0}{\partial x}+V_0\frac{\partial V_0}{\partial y}=-\lambda g\frac{\partial \zeta}{\partial y}-(1-\lambda)g\frac{\partial \zeta_0}{\partial y}+\frac{\tau_{iy}-\tau_{by}}{\rho_0 h_0}-fU_0 \tag{2.105}$$

where

$$\frac{\tau_{sx}}{\rho_0} = k_w W_x \sqrt{(W_x^2 + W_y^2)}$$
$$\frac{\tau_{sy}}{\rho_0} = k_w W_y \sqrt{(W_x^2 + W_y^2)} \qquad (2.106)$$

$$\frac{\tau_{ix}}{\rho_0} = (U - U_0)k_i \sqrt{[(U - U_0)^2 + (V - V_0)^2]}$$
$$\frac{\tau_{iy}}{\rho_0} = (V - V_0)k_i \sqrt{[(U - U_0)^2 + (V - V_0)^2]} \qquad (2.107)$$

$$\frac{\tau_{bx}}{\rho_0} = k_b U_0 \sqrt{(U_0^2 + V_0^2)}$$
$$\frac{\tau_{by}}{\rho_0} = k_b V_0 \sqrt{(U_0^2 + V_0^2)} \qquad (2.108)$$

The coefficients k_w (wind friction), k_i (interface friction), k_b (bed friction) have values depending on the wind, flow, fluid and bed morphology. An approximation to their orders of magnitude is, $O[k_w] = 10^{-6}$, $O[k_i] = 10^{-3}$, $O[k_b] = 10^{-2}$.

The above 2-layer model can describe the circulation in a stratified medium due

(1) to the wind influence on the free surface
(2) to incident long waves through the open sea boundary or induced from barometric pressure fluctuations
(3) to primary or secondary interface gradients (baroclinic flows, internal waves). It is completed by appropriate boundary conditions completely analogous to those for the homogeneous medium. At the open sea boundary the free radiation conditions for the two layers take the form:

Lower layer: $$U_{n_0} h_0 = -\zeta_0 \sqrt{\left[\frac{g h h_0 (1 - \lambda)}{(h + h_0)}\right]} \qquad (2.109)$$

Upper layer: $$U_n h = -\zeta \sqrt{[g(h + h_0)]} \qquad (2.110)$$

The numerical solution of the 2-layer model follows the same methodology as the 2DH model for a homogeneous fluid, i.e. an explicit finite difference scheme on a staggered grid. Program 14 gives the structure for the one-dimensional case.

PROGRAM 14: 1-D STRATIFIED FLOW MODEL, TWO-LAYER

```
  5REM 1-D STRATIFIED FLOW MODEL 2LAYERS
 10DIMU(20),UN(20),UO(20),UON(20),H(20),Z(20),HO
(20),ZO(20),DX(20)
 20READ DT,CF,CI,IM,EL,NM,PER,EDH,W,CS
 25DATA...
 30FORI=0TOIM:READH(I),HO(I),DX(I):NEXTI
 40DATA...
 50FORI=1TOIM:U(I)=0:UN(I)=0:UO(I)=0:UON(I)=0:NE
XTI
 60TSXO=CS*W*ABS(W):N=0:T=0
 70N=N+1:T=T+DT
 80IFN<PER/4 THENTSX=TSXO*SIN(2*PI*N/PER) ELSE T
SX=TSXO
 90FORI=1TOIM-1:PT=ZO(I):ZO(1)=ZO(I)-DT/DX(I)/2*
((HO(I+1)+HO(1))*UO(I+1)-(HO(I)+HO(I-1))*UO(I)):Z(
I)=Z(I)+ZO(I)-PT-DT/DX(I)/2*((H(I)+H(I+1))*U(I+1)-
(H(I)+H(I-1))*U(I)):NEXTI
 100FORI=2TOIM-1:HM=(H(I)+H(I-1))/2:HMO=(HO(I)+HO
(I-1))/2:TBX=UO(I)*ABS(UO(I))*CF:DXM=(DX(I)+DX(I-1
))/2:TIX=CI*(U(I)-UO(I))*ABS(U(I)-UO(I))
 110UN(I)=U(I)*EDH+(U(I+1)+U(I-1))*(1-EDH)/2+DT*(
-((U(I+1)+U(I))^2-(U(I)+U(I-1))^2)/8/DXM-9.81*(Z(I
)-Z(I-1))/DXM-(TIX-TSX)/EL/HM)
 120UON(I)=UO(I)*EDH+(UO(I+1)+UO(I-1))*(1-EDH)/2+
DT*(-((UO(I+1)+UO(I))^2-(UO(I)+UO(I-1))^2)/8/DXM-9
.81*(ZO(I)-ZO(I-1))/DXM*(1-EL)-9.81*EL*(Z(I)-Z(I-1
))/DXM-(TBX-TIX)/HMO) :NEXTI
 130UN(1)=-Z(1)*SQR(9.81*(H(1)+HO(1)))/H(1):UON(1
)=-ZO(1)*SQR(9.81*H(1)*HO(1)/(H(1)+HO(1))*(1-EL))/
HO(1)
 140FORI=1TOIM:U(I)=UN(I):UO(I)=UON(I):NEXTI
 150IFN/50<>INT(N/50)THEN GOTO70
 160EK=0:EKO=0:FORI=1TOIM-1:EK=EK+((U(I)+U(I+1))/
2)^2*H(I):EKO=EKO+((UO(I)+UO(I+1))/2)^2*HO(I):NEXT
I
 170PRINTT,EK,EKO,Z(6),ZO(6),U(4),UO(4):IF(N<NM)
THEN GOTO70
 180PRINT"Z":PRINT:FORI=1TOIM :PRINTZ(I);:NEXTI:P
RINT:PRINT"ZO":FORI=1TOIM:PRINTZO(I);:NEXTI:PRINT:
PRINT"U":FORI=1TOIM:PRINTU(I);:NEXTI
 190PRINT:PRINT"UO":FORI=1TOIM:PRINTUO(I);:NEXTI
:IF(N<NM) THEN GOTO70
 200END
```

Description of variables:

DT, DX = time and space discretisation steps
CF = bed friction coefficient
CI = interface friction coefficient

CS = surface wind friction coefficient
IM = number of cross-sections for the discretisation of the flow field
NM = number of time steps in the numerical solution
EL = density ratio λ
EDH = weight factor in the Lax type time difference
PER = time for the sinusoidal increase of wind intensity from a cold start to steady blow
W = wind velocity
H, H0 = initial depths in no-flow conditions for the two layers (horizontal interface and surface).

The program refers to a linear basin (the left and right boundaries

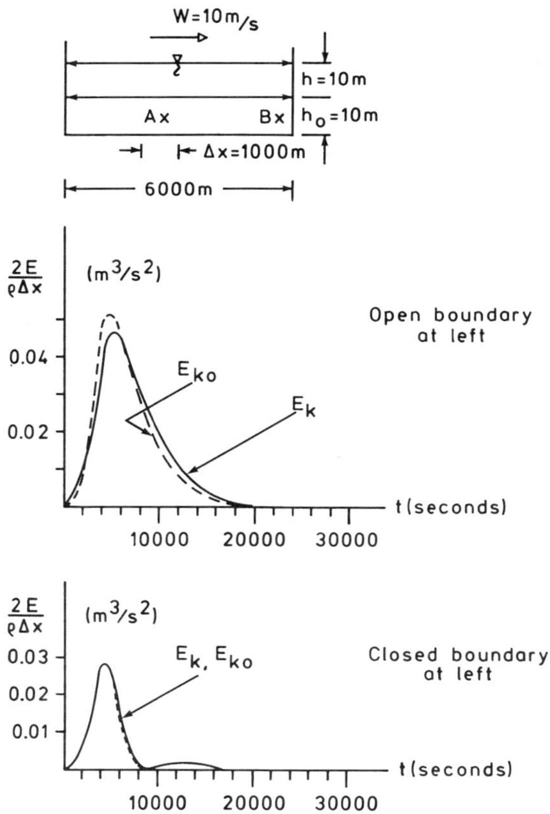

Fig. 2.24 Flow domain and evolution of kinetic energy for enclosed and open domains in 1-D stratified flow model

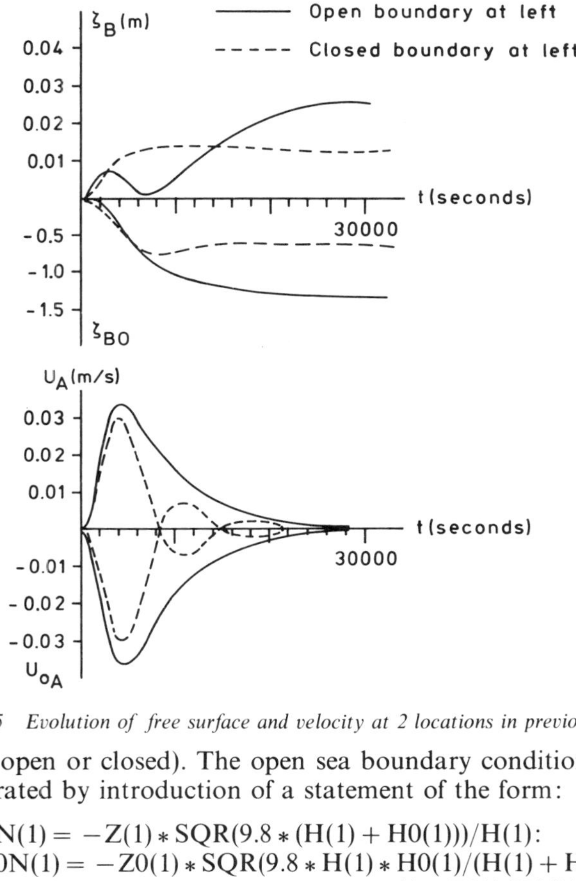

Fig. 2.25 Evolution of free surface and velocity at 2 locations in previous model

may be open or closed). The open sea boundary condition can be incorporated by introduction of a statement of the form:

```
130 UN(1) = -Z(1)*SQR(9.8*(H(1)+H0(1)))/H(1):
    U0N(1) = -Z0(1)*SQR(9.8*H(1)*H0(1)/(H(1)+H0(1))
                                    *(1-EL))/H0(1)
```

The application refers to the description of the transient flow in a flow domain extending over 6 km with two layers of equal depth, $h = h_0 = 10$ m, under the influence of a wind of 10 m/s. Two cases are examined, a laterally closed channel and a channel with an open sea boundary at left. The program data are: DT = 20 s, CF = 0.05, CI = 0.005, IM = 7, EL = 0.98, (fresh water above sea water), NM = 3000, PR = 800 s, EDH = 0.98, W = 10 m/s, CS = 0.000,005, H(I) = 10 m, H0(I) = 10 m (I = 1–7), DX(I) = 1000 m.

Figure 2.24 shows the evolution of the kinetic energies of the two layers for closed and open flow fields. Figure 2.25 shows the transient evolution up to steady conditions of the ζ, ζ_0 on the right closed boundary and the u, u_0 in the middle of the flow field.

The comparison leads to the following preliminary conclusions:

(1) The transients develop differently for the two cases of open and closed flow domains. The interfacial waves dominate the phenomenon.
(2) The expected difference in the orders of magnitude of ζ and ζ_0 are revealed. Under steady no-flow equilibrium conditions their ratio reaches the value of $\approx 1/(1-\lambda) = 50$.

3

Transport of pollutants in coastal regions

3.1 THE GENERAL MATHEMATICAL MODEL FOR TURBULENT ADVECTIVE DIFFUSION

3.1.1 Insight into the physical phenomenon

The transport of pollution in nature is done mainly by fluids. The coastal regions were and are still used as the final receptor of local and land based pollution sources (oil slicks, domestic sewage, industrial waste).

Present day engineers have become very sensitive to and active in problems of coastal environment protection and consequently this chapter is devoted to the formulation of mathematical models describing the phenomenon of advection and diffusion of pollutants and to the presentation of various numerical solutions. Initially, some specific aspects of the physical processes involved will be presented.

Similar to the case of momentum transport in coastal geophysical domains, in the case of mass transport, the assumption of nearly horizontal flow is valid away from the source of pollution. In the near field the phenomenon may have a fully three-dimensional nature (e.g., a buoyant plume developing from a submarine sewage outfall) but in the far field the horizontal motion prevails over the vertical. In the far field the phenomenon can become two-dimensional, in a horizontal sense, and two extreme cases may be distinguished.

In the first case, due to intense vertical mixing, adequate homogenisation of the pollutant along the depth is achieved (or assumed) and the theory of one- or two-dimensional dispersion is realistically applicable.

In the second case, the pollutant, mainly due to strong buoyancy effects (oil in water, or sewage in stratified environment), is concentrated in a thin layer of specific known depth (surface or interface) and the phenomenon may be considered as evolving in the 2-D space xOy. In the first case the model has to be written with respect to the depth-mean concentration $C(x, y, t)$ while in the second case with respect to a local concentration $c(x, y, t)$, referring to a specific zone at depth z and specific known thickness.

It has already been mentioned that pollution transport is accomplished by two distinct physical mechanisms, advection and diffusion. The first describes the entrainment of the pollutant material in the ambient flow and its transport either as solution or as suspension by the fluid, with velocity equal to the fluid velocity (assuming negligible resistance to the flow).

The second mechanism describes, in the case of laminar flow, the Brownian motion of the pollutant's molecules, resulting in the continuous increase of the area that the pollutant, originating from a local source, occupies. In the case of turbulent flow, the motions at molecular level become negligible before the similar process of entraining pollutant masses in turbulent eddies, leading to the same result but at a much faster rate.

The rate of diffusion even in the case of turbulent flow can be described by a diffusion coefficient, according to the Boussinesq approximation. This is the eddy diffusion coefficient, analogous to the eddy viscosity coefficient and some orders of magnitude greater than the molecular diffusion coefficient, a function, in general, of the hydrodynamics of the pollutant-carrying fluid. It will be symbolised by D (m^2/s).

The proportion by which the two mechanisms of advection and diffusion contribute to the final pollutant transport is quantified through the Péclet number.

$$P_e = UL/D \tag{3.1}$$

where U (m/s) is the characteristic velocity of the transporting fluid, L (m), is a characteristic length of the flow domain and D (m^2/s), the diffusion coefficient. Values of P_e, $O[P_e] > 1$ demonstrate that the advection dominates over the fiffusion.

Finally, in the analysis of the physical processes involved in the transport of pollutants, the distinction between conservative and non-conservative pollutants must be made. If the total mass of the advected-diffused pollutant is conserved during its transport, then it is characterised as a conservative one. Otherwise, if its discharge in the carrying fluid initiates physical, chemical or biological processes that result in a decrease or increase in its initial mass, then it is characterised as a non-conservative one. Such processes are:

(1) Biological increase or decrease of bacteria.
(2) Chemical reaction with the environment (oxidation, UV destabilisation, etc.).
(3) Settlement of flocculated pollutant molecules.

3.1.2 Formulation of the mathematical model

The mathematical model for pollution transport is written with respect to the unknown scalar function of the concentration of pollutant $c(x, y, z, t)$. The concentration is measured either as the volume of the pollutant, or as pollutant units in unit volume of the fluid (examples of units are, ppm, mg per litre, number of *Escherichia coli* per litre). The model consists of one equation describing the conservation of the mass of pollutant during its motion. Fick's law and a Boussinesq approximation are assumed to be valid in the case of turbulent flow. Examining the balance of the inflowing–outflowing mass from two cross-sections in a simple one-dimensional flow with velocity u, it is found that:

$$\frac{\mathrm{d}c}{\mathrm{d}t}=\frac{\partial c}{\partial t}+\frac{\partial(uc)}{\partial x}=\frac{\partial}{\partial x}\left(D\frac{\partial c}{\partial x}\right)-\lambda c \tag{3.2}$$

where D is the eddy diffusion coefficient and λ (s^{-1}) the decay coefficient. In the most general case of three-dimensional pollutant transport, Equation (3.2) becomes

$$\frac{\partial c}{\partial t}+\frac{\partial(cu)}{\partial x}+\frac{\partial(cv)}{\partial y}+\frac{\partial(cw)}{\partial z}=\frac{\partial}{\partial x}\left(D_x\frac{\partial c}{\partial x}\right)+\frac{\partial}{\partial y}\left(D_y\frac{\partial c}{\partial y}\right)+\frac{\partial}{\partial z}\left(D_z\frac{\partial c}{\partial z}\right)-\lambda c \tag{3.3}$$

In the case of coastal pollution, the specific 3-D form of (3.3) becomes

$$\frac{\partial c}{\partial t}+\frac{\partial(cu)}{\partial x}+\frac{\partial(cv)}{\partial y}=\frac{\partial}{\partial x}\left(D_h\frac{\partial c}{\partial x}\right)+\frac{\partial}{\partial y}\left(D_h\frac{\partial c}{\partial y}\right)+\frac{\partial}{\partial z}\left(D_v\frac{\partial c}{\partial z}\right)-\lambda c \tag{3.3a}$$

and in the 2-D depth averaged transport or transport in a horizontal layer, (3.3) becomes

$$\frac{\partial C}{\partial t}+\frac{\partial(CU)}{\partial x}+\frac{\partial(CV)}{\partial y}=\frac{\partial}{\partial x}\left(R_x\frac{\partial C}{\partial x}\right)+\frac{\partial}{\partial y}\left(R_y\frac{\partial C}{\partial y}\right)-\lambda C \tag{3.3b}$$

where R_x, R are dispersion coefficients and U, V are depth mean velocities.

The boundary conditions completing the model can be of the following form:

(1) Solid boundaries: The zero normal flux is expressed as

$$\frac{\partial c}{\partial n}=0 \tag{3.4}$$

(2) Free transmission boundaries: Such as open sea boundaries or any pollution trap permitting inflow of mass without back return. Here, a simple condition of uniform flux applies, i.e.

$$\frac{\partial}{\partial n}\left(D\frac{\partial c}{\partial n}\right)=0 \qquad \text{or} \qquad \frac{\partial^2 c}{\partial n^2}=0 \qquad \text{for } D\simeq\text{const.} \tag{3.5}$$

(3) Pollution sources. Here, the pollutant concentration is known (prefixed).

3.2 NUMERICAL SOLUTIONS. CRITICAL PRESENTATION OF FINITE DIFFERENCE SCHEMES

From the field Equation (3.3) of the mathematical model for the turbulent advective diffusion of a conservative or non-conservative pollutant it can be shown that the mathematical problem is the solution of a linear PDE of mixed type (hyperbolic + parabolic). The advection is accomplished with celerity equal to the fluid velocity (u, v) while the diffusion is done at a rate analogous to the assumed D values. According to the Péclet values, the hyperbolic or the parabolic part prevails. It can be said a priori that the major problem in the numerical solution of the model is control of the numerical errors intruding in the solution in the form of numerical diffusion or dispersion. The explicit forms of finite difference schemes are subject to stability criteria regarding the space and time discretisation steps $\Delta x, \Delta t$. They are mathematically expressed by

$$u_{\text{max}}\frac{\Delta t}{\Delta x}<1, \qquad D_{\text{max}}\frac{\Delta t}{\Delta x^2}<\frac{1}{2} \tag{3.6}$$

while the implicit forms are numerically stable for all $\Delta x, \Delta t$. The problem in this case is the loss of information with the increase of the values of the expression in (3.6) far beyond those limits.

Let us consider the one-dimensional case with constant u, D, to demonstrate the properties of various solution schemes,

$$\frac{\partial c}{\partial t}+u\frac{\partial c}{\partial x}=D\frac{\partial^2 c}{\partial x^2} \tag{3.7}$$

The parabolic part of the differential operator is approximated by centered finite differences, either explicit or implicit (Crank–Nicolson)

$$\frac{\partial c}{\partial t}=D\frac{\partial^2 c}{\partial x^2}\rightarrow\frac{c_i^{n+1}-c_i^n}{\Delta t}=\frac{D}{\Delta x^2}(c_{i+1}^n-2c_i^n+c_{i-1}^n) \tag{3.9}$$

or
$$\frac{c_i^{n+1} - c_i^n}{\Delta t} = \frac{D}{\Delta x^2}(c_{i+1}^{n+\frac{1}{2}} - 2c_i^{n+\frac{1}{2}} + c_{i-1}^{n+\frac{1}{2}}) \qquad (3.10)$$

where $c_i^{n+\frac{1}{2}} = (c_i^n + c_i^{n+1})/2$. The numerical solution of (3.9) is subject to the second inequality constraint of (3.6) while (3.10) is unconditionally stable. Experiments have shown that the numerical solutions obtained by (3.9) or (3.10) behave smoothly and no serious errors creep in. As was already mentioned in Section 3.1 the most difficult task is the approximation of the advective hyperbolic part,

$$\frac{\partial c}{\partial t} + u\frac{\partial c}{\partial x} = 0 \qquad (3.11)$$

The most common easily used and practical numerical schemes for the approximation of (3.11) are presented below, although more sophisticated and effective ones are continuously being investigated.

(1) Backward (upwind) differences

$$\frac{c_i^{n+1} - c_i^n}{\Delta t} = -u\frac{c_i^n - c_{i-1}^n}{\Delta x} \qquad (3.12)$$

(2) Lax dissipative scheme

$$\frac{\theta c_i^{n+1} - \dfrac{1-\theta}{2}(c_{i+1}^n + c_{i-1}^n)}{\Delta t} = \frac{-u}{2\Delta x}(c_{i+1}^n - c_{i-1}^n) \qquad (3.13)$$

(3) Lax–Wendroff scheme

$$\frac{c_i^{n+1} - c_i^n}{\Delta t} = \frac{-u}{2\Delta x}(c_{i+1}^n - c_{i-1}^n) + \frac{u^2\Delta t}{\Delta x^2}(c_{i+1}^n - 2c_i^n + c_{i-1}^n) \qquad (3.14)$$

(4) Leap-frog scheme

$$\frac{c_i^{n+1} - c_i^{n-1}}{2\Delta t} = \frac{-u}{2\Delta x}(c_{i+1}^n - c_{i-1}^n) \qquad (3.15)$$

(5) Fully implicit scheme

$$\frac{c_i^{n+1} - c_i^n}{\Delta t} = \frac{-u}{4\Delta x}(c_{i+1}^{n+1} + c_{i+1}^n - c_{i-1}^{n+1} - c_{i-1}^n) \qquad (3.16)$$

(6) McCormack's scheme

$$\tilde{c}_i^n = c_i^n - \frac{u\Delta t}{\Delta x}(c_{i+1}^n - c_i^n) \qquad (3.17)$$

$$c_i^{n+1} = \tfrac{1}{2}(\tilde{c}_i^n + c_i^n) - \frac{u\Delta t}{\Delta x}(\tilde{c}_i^n - \tilde{c}_{i-1}^n) \qquad (3.18)$$

$$c_i^{n+1} = c_i^n - \frac{u\Delta t}{2\Delta x}(c_{i+1}^n + c_i^n - 2c_{i-1}^n) + \left(\frac{u\Delta t}{\Delta x}\right)^2 (c_{i+1}^n - 2c_i^n + c_{i-1}^n) \qquad (3.19)$$

(7) Fromm's scheme

$$c_i^{n+1} = c_i^n - \frac{u\Delta t}{4\Delta x}(c_{i+1}^n - c_{i-1}^n + c_i^n - c_{i-2}^n) + \left(\frac{u\Delta t}{2\Delta x}\right)^2 (c_{i+1}^n - 2c_i^n + c_{i-1}^n) + \frac{u^2 \cdot \Delta t^2 - 2u \cdot \Delta t \cdot \Delta x}{4\Delta x^2}(c_{i-2}^n - 2c_{i-1}^n + c_i^n) \qquad (3.20)$$

(8) Generalised box explicit scheme

$$\left(\frac{c_i^{n+1} - c_i^n}{\Delta t}\right)(1-\theta) + \theta\left(\frac{c_{i-1}^{n+1} - c_{i-1}^n}{\Delta t}\right) = \frac{-u}{2\Delta x}(c_i^{n+1} - c_{i-1}^{n+1} + c_i^n - c_{i-1}^n) \qquad (3.21)$$

The comparative numerical experiments were performed with $\Delta x = 1$, $\Delta t = 0.8$, $u = 1$ (Courant number, $u\Delta t/\Delta x = 0.8$). The initial condition was $c(x, 0) = 0$ and the upstream boundary condition was $c(0, t) = 1$. The analytical solution has the form of the Heaviside function with the step moving in the positive x direction with velocity $= u$. The numerical solution of the nine abovementioned schemes were programmed in BASIC. Figure 3.1 contains the concentration distributions after some time compared to the analytical solution.

A BASIC listing for the Fromm scheme is given in Program 15.

PROGRAM 15: 1-D ADVECTIVE POLLUTANT TRANSPORT MODEL—FROMM SCHEME

```
    5REM 1-D ADVECTIVE POLLUTANT TRANSPORT MODEL F
ROMM SCHEME
   10DIM C(55),CN(55)
   20READ DT,DX,U,IM,NM
   30DATA....
   40C(1)=1:CN(1)=1:C(0)=1
   50N=0:T=0
   60N=N+1:T=T+DT
   70FORI=2TOIM-1:CN(I)=C(I)-U*DT/4/DX*(C(I+1)-C(I
```

```
-1)+C(I)-C(I-2))+(U*DT/DX/2)^2*(C(I+1)-2*C(I)+C(I-
1))+(U^2*DT^2-2*U*DT*DX)/4/DX^2*(C(I-2)-2*C(I-1)+C
(I)):NEXTI
    80CN(IM)=2*CN(IM-1)-CN(IM-2)
    85FORI=1TOIM:C(I)=CN(I):NEXTI
    90IFN/10<>INT(N/10) THEN GOTO 60
   100PRINT:PRINTT:FORI=1TOIM:PRINTC(I);:NEXTI
   110IFN<NM THEN GOTO 60
   120END
  1000PRINTN:MOVE 0,C(1)*200+500:FORI=2TOIM:DRAW(I-
1)*20,C(I)*200+500:NEXTI
```

Variable description:

DT = time step
DX = space step
U = velocity
IM = number of computation points
C(I) = concentration value at point $\#I$.

From Fig. 3.1 it is understood that the main deficiency in the numerical solution for the purely advection model is the numerical diffusion. This is minimal in the Fromm and Box schemes. Others, such as the Lax–Wendroff, leap-frog and implicit schemes contain also some numerical dispersion. The full form of the advective diffusion model is numerically treated through the combined use of the finite difference schemes presented above for parabolic and 1st order hyperbolic operators.

The application of the traditional finite difference schemes described leads to the general conclusion that the numerical solutions are adequately accurate (from an engineering point of view) for Péclet numbers less than 10. The numerical diffusion for $P_e > 10$ makes most of the above schemes inefficient. Contemporary research both in the finite difference and finite element directions is oriented mainly towards upwind schemes and dissipative schemes with controllable numerical diffusion.

The extension of the model to two dimensions introduces no further difficulty. The field discretisation is performed along the Ox and Oy axes with steps $\Delta x, \Delta y$ and the numerical computation of the unknown concentration function requires a two-dimensional matrix $C(I, J) \rightarrow c_{ij}$ (where $x_i = i\Delta x$, $y_i = j\Delta y$). Explicit or implicit schemes like the ones described above can be used in this case. The most simple explicit scheme contains forward differences for the time derivative and centered differences for the space derivatives. The inductive formula for the computation of concentration at time level

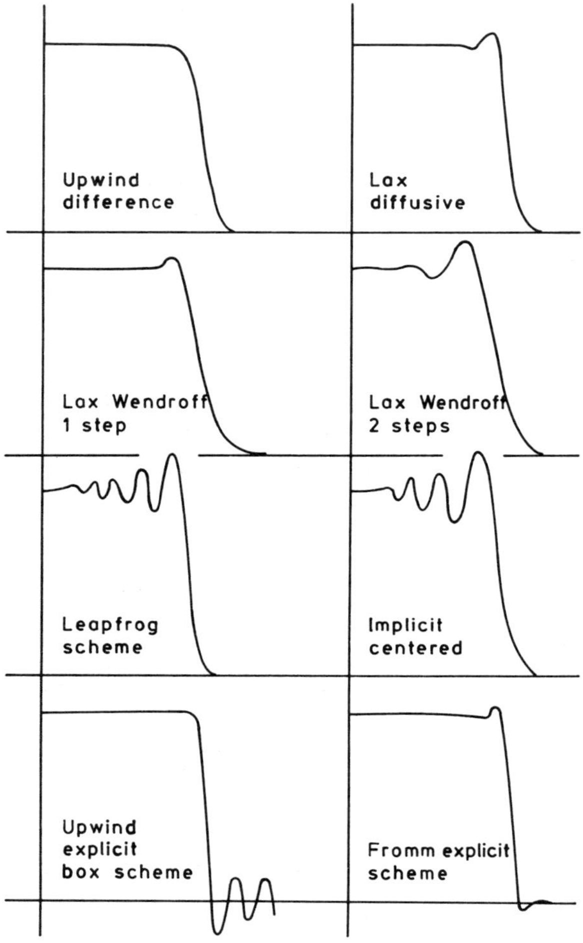

Fig. 3.1 1-D advection equation. Solutions by various finite difference schemes

$n+1$, at a point i, j of the flow domain, has the form

$$c_{ij}^{n+1} = c_{ij}^n - \frac{(\Delta t)u}{2\Delta x}(c_{i+1j}^n - c_{i-1j}^n) - \frac{(\Delta t)v}{2\Delta y}(c_{ij+1}^n - c_{ij-1}^n)$$
$$+ \frac{(\Delta t)D_x}{\Delta x^2}(c_{i+1j}^n - 2c_{ij}^n + c_{i-1j}^n)$$
$$+ \frac{(\Delta t)D_y}{\Delta y^2}(c_{ij+1}^n - 2c_{ij}^n + c_{ij-1}^n) \qquad (3.22)$$

(It should be noted that centered differences for pure advection are unstable and the diffusion part of the operator is necessary to suppress instability.) The explicit scheme is very easily programmable but the time step Δt is limited by the known stability criteria. The use of implicit finite differences for both parts of the operators $u\,\partial c/\partial x$, $D\,\partial^2 c/\partial x^2$ lead to an algebraic system in IM × JM unknowns (where IM, JM are the maximum values of the indices I, J along x, y respectively). The fully implicit scheme leads to the inductive relation

$$\frac{c_{ij}^{n+1}-c_{ij}^{n}}{\Delta t} = -u\left(\frac{c_{i+1j}^{n+1}-c_{i-1j}^{n+1}}{2\Delta x}\right)-v\left(\frac{c_{ij+1}^{n+1}-c_{ij-1}^{n+1}}{2\Delta y}\right) + \frac{D_x(c_{i+1j}^{n+1}-2c_{ij}^{n+1}+c_{i-1j}^{n+1})}{\Delta x^2} + \frac{D_y(c_{ij+1}^{n+1}-2c_{ij}^{n+1}+c_{ij+1}^{n+1})}{\Delta y^2} \tag{3.23}$$

In order to avoid the large computational volume required for the fully implicit scheme (retaining at the same time the qualifications of the implicitness) it is preferable to apply the ADI technique (Alternating Directions Implicit). According to this method the solution advances from time level n to $n+1$ using an implicit scheme along y and explicit along x, while alternately from $n+1$ to $n+2$, using an implicit scheme along x and explicit along y. The inductive formula takes the forms:

$$c_{ij}^{n+1} = c_{ij}^{n} - \frac{u\Delta t}{2\Delta x}(c_{i+1j}^{n}-c_{i-1j}^{n}) - \frac{v\Delta t}{2\Delta y}(c_{ij+1}^{n+1}-c_{ij-1}^{n+1}) - \frac{(\Delta t)D_x}{\Delta x^2}(c_{i+1j}^{n}-2c_{ij}^{n}+c_{i-1j}^{n}) + \frac{(\Delta t)D_y}{\Delta y^2}(c_{ij+1}^{n+\frac{1}{2}}-2c_{ij}^{n+\frac{1}{2}}+c_{ij-1}^{n+\frac{1}{2}}) \tag{3.24}$$

$$c_{ij}^{n+2} = c_{ij}^{n+1} - \frac{u\Delta t}{2\Delta x}(c_{i+1j}^{n+2}-c_{i-1j}^{n+2}) - \frac{v\Delta t}{2\Delta y}(c_{ij+1}^{n+1}-c_{ij-1}^{n+1}) - \frac{(\Delta t)D_x}{\Delta x^2}(c_{i+1j}^{n+\frac{3}{2}}-2c_{ij}^{n+\frac{3}{2}}+c_{i-1j}^{n+\frac{3}{2}}) + \frac{(\Delta t)D_y}{\Delta y^2}(c_{ij+1}^{n+1}-2c_{ij}^{n+1}+c_{ij-1}^{n+1}) \tag{3.25}$$

This technique leads to the solution of IM algebraic systems in JM unknowns when advancing from n to $n+1$ and to the solution of JM systems in IM unknowns when advancing from $n+1$ to $n+2$ time levels.

The numerical solution algorithm organised in BASIC is given in Program 16. It refers to an orthogonal flow domain with $\Delta x = \Delta y$, a point source of pollution (c = constant) and perimetric boundary conditions, $\partial^2 c/\partial n^2 = 0$. The deriving algebraic systems have tridiagonal matrices and their solution is performed using the Thomas algorithm.

PROGRAM 16: 2-D ADVECTIVE DIFFUSION ADI METHOD. THOMAS'S SOLUTION

```
    5 REM 2-D ADVECTIVE DIFFUSION ADI MODEL THOMAS
 SOLN
   10DIMC(20,20),CN(20,20),AI(20),BI(20),CI(20),FI
(20),AJ(20),BJ(20),CJ(20),FJ(20),A(20),B(20),CO(20
),F(20),X(20),AT(20),BT(20),FT(20),CT(20)
   20READDT,DX,U,V,K,IM,JM,NM,IP,JP
   30DATA...
   40A0=1/DT+2*K/DX^2:A1=-V/2/DX+K/2/DX^2:A2=V/2/D
X+K/2/DX^2:A3=1/DT-2*K/DX^2:A4=K/DX^2-U/2/DX:A5=K/
DX^2+U/2/DX:A6=-U/2/DX+K/2/DX^2:A7=U/2/DX+K/2/DX^2
:A8=-V/2/DX+K/DX^2:A9=V/2/DX+K/DX^2:A10=K/2/DX^2
   42FORJ=2TOJM-1:AJ(J)=-A1-A10:BJ(J)=A0:CJ(J)=-A2
-A10:NEXTJ:BJ(2)=BJ(2)+2*CJ(2):AJ(2)=AJ(2)-CJ(2):B
J(JM-1)=BJ(JM-1)+2*AJ(JM-1):CJ(JM-1)=CJ(JM-1)-AJ(J
M-1)
   44FORI=2TOIM-1:AI(I)=-A6-A10:BI(I)=A0:CI(I)=-A7
-A10:NEXTI:BI(2)=BI(2)+2*CI(2):AI(2)=AI(2)-CI(2):B
I(IM-1)=BI(IM-1)+2*AI(IM-1):CI(IM-1)=CI(IM-1)-AI(I
M-1)
   50T=0:N=0
   60T=T+DT:N=N+1
   70IFN/2=INT(N/2) THEN GOTO80 ELSE GOTO200
   80FORI=2TOIM-1
   90FORJ=2TOJM-1:FJ(J)=C(I.J)*A3+C(I+1,J)*A4+C(I-
1,J)*A5:NEXTJ
  100FORJ=2TOJM-1:A(J)=AJ(J):B(J)=BJ(J):CO(J)=CJ(J
):F(J)=FJ(J):NEXTJ:KM=JM-1
  110GOSUB1000
  120FORJ=2TOJM-1:CN(I,J)=X(J):NEXTJ:CN(I,1)=CN(I,
2)*2-CN(I,3):CN(I,JM)=CN(I,JM-1)*2-CN(I,JM-2)
  180NEXTI
  190FORJ=1TOJM:CN(1,J)=2*CN(2,J)-CN(3,J):CN(IM,J)
=2*CN(IM-1,J)-CN(IM-2,J):NEXTJ
  195GOTO305
  200FORJ=2TOJM-1
  210FORI=2TOIM-1:FI(I)=C(I,J)*A3+C(I,J+1)*A8+C(I,
J-1)*A9:NEXTI
```

```
220FORI=2TOIM-1:A(I)=AI(I):B(I)=BI(I):CO(I)=CI(I
):F(I)=FI(I):NEXTI:KM=IM-1
230GOSUB1000
240FORI=2TOIM-1:CN(I,J)=X(I):NEXTI:CN(1,J)=CN(2,
J)*2-CN(3,J):CN(IM,J)=CN(IM-1,J)*2-CN(IM-2,J)
250NEXTJ
300FORI=1TOIM:CN(I,1)=2*CN(I,2)-CN(I,3):CN(I,JM)
=2*CN(I,JM-1)-CN(I,JM-2):NEXTI
305CN(IP,JP)=1
310FORI=1TOIM:FORJ=1TOJM:C(I,J)=CN(I,J):NEXTJ:NE
XTI
320PRINT:PRINTT:PRINT:FORJ=JM TO 1STEP-1:FORI=1T
OIM:PRINTC(I,J);:NEXTI:NEXTJ
330IFN<NM THEN GOTO60
340END
1000BT(2)=B(2):AT(2)=A(2):FT(2)=F(2)
1010FORK=3TOKM:BT(K)=B(K)*BT(K-1)-AT(K-1)*CO(K):A
T(K)=A(K)*BT(K-1):FT(K)=F(K)*BT(K-1)-FT(K-1)*CO(K)
:NEXTK
1020X(KM)=FT(KM)/BT(KM):FORK=KM-1 TO 2 STEP-1:X(K
)=(FT(K)-AT(K)*X(K+1))/BT(K):NEXTK
1030RETURN
```

Description of variables:

DT, DX = time and space discretisation steps
IM, JM = range of indices i, j along Ox, Oy
U, V = velocity components in x, y directions (uniform velocity field is used in this example)
K = constant diffusion coefficient
IP, JP = position indices for the pollution source ($c = 1$)
NM = number of time steps.

From sections *a–a*, *b–b* of Fig. 3.2 it can be seen that the alternating implicit and explicit solutions result in a slight oscillation in the values of concentrations between successive time steps.

In concluding the various finite difference schemes it can be stated that the numerical description of the propagation of a linear pollution wave with simultaneous diffusion is subject to numerical diffusion and dispersion errors that sometimes mask the solution. The solutions refer to the far field and no information can be obtained in the subgrid scale. Current research is striving to overcome those problems with

(1) the use of finite elements or nested finite difference grids for the more detailed description of the c's in areas of interest
(2) the use of upwind finite differences or upwind biased base

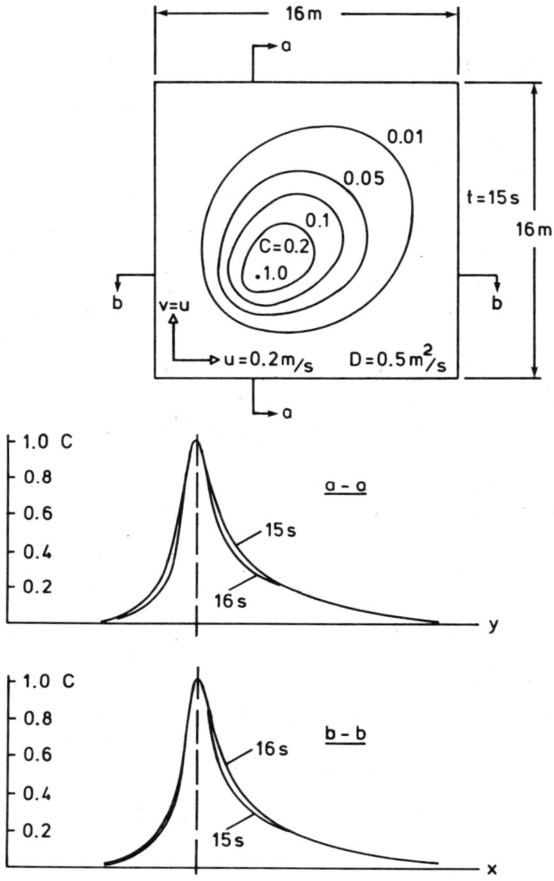

Fig. 3.2 2-D advective diffusion. Solution by ADI method

functions in finite elements characterised by minimal numerical diffusion

(3) the description of the transport process in a Lagrangian frame.

This last technique permits the tracking of the pollution both in the near and the far field. A method for the description of both the advection and diffusion processes in a Lagrangian frame, known as the tracer technique, is given in detail in Section 3.3.

3.3 THE TRACER TECHNIQUE

3.3.1 Principle and general description of the method

The distinction between the flow description in Eulerian and Lagrangian frame of reference is assumed known. The same distinction can be extended to the description of the transport of a substance diluted or suspended in a carrying fluid. Instead of describing the concentration (number of solid suspended particles or diluted matter) of the substance at various fixed points it is theoretically possible to track the motion of all the 'particles' of the substance as they are carried by the fluid. The Lagrangian description requires the continuous tracking of all the particles, into which the suspended or diluted pollutant can be discretised, in order to compute the local concentration as the limit $\delta m/\delta\Omega$ where δm is the pollutant mass and $\delta\Omega$ the volume of the dilution. These solid particles are infinite in reality but they can be adequately approximated by a large but finite number. By the word large we mean a number that is compatible with the computer facility available.

As was described above during the advective diffusion of a non-conservative pollutant, three processes are superimposed, that of advection (with velocity equal to the velocity of the fluid, assuming that the suspended or diluted matter does not react on the hydrodynamic properties of its environment), that of molecular or turbulent diffusion, and that of decomposition and disappearance of the substance. In the Lagrangian description these three processes are simulated in the following manner:

(1) Advection is accomplished by the local fluid velocity. From the coordinates of the tracked particle (the tracer) the local fluid velocity is computed (possibly interpolated from neighbouring locations) and the coordinates of the particle after time Δt are computed from simple kinematics.
(2) Decomposition, in the case of a non-conservative substance, is simulated by the withdrawal from the flow domain of a number of particles at random (that number is related to the decay factor λ).
(3) Diffusion is simulated, according to the subsequent detailed description, by a random Brownian motion of the particles. Correspondence of random motion and diffusion is physically realistic; its mathematical quantification was demonstrated by Einstein at the beginning of the century.

The simulation of these three processes permits the tracking of the large number of discrete particles evolving that approximates the pollutant. If the concentration at a certain location is required, it can be computed from the number of particles that are contained in a fixed space (for example, the grid mesh $\Delta x \cdot \Delta y$) by comparison with the number of particles that are included in a mesh with known concentration. For example, if it is known that 20 particles in a mesh correspond to 10 ppm then 5 particles in another mesh correspond to a concentration of 2.5 ppm.

From the description of the tracer technique it is understood that the only process that requires further analysis is the simulation of diffusion. This is presented in detail in Section 3.3.2.

3.3.2 Numerical simulation of diffusion

Let us assume the case of one-dimensional diffusion described by the equation

$$\frac{\partial c}{\partial t} = D\frac{\partial^2 c}{\partial x^2} \tag{3.26}$$

For the initial condition $c(x,0) = \delta(x - x_0)$, that is an initial unit mass of pollutant concentrated at $x = x_0$, the solution $c(x, t)$ has the form of a Gaussian bell and there is a relation between the variance of that distribution and the diffusion coefficient

$$\sigma_x^2 = 2Dt \tag{3.27}$$

Let us assume instead a large number of particles originating from the same origin at $t = 0$ and executing random walks of length Δx (following the binomial distribution) in time Δt. According to the central limit theorem, on average, their position along the x axis follows a normal distribution with variance

$$\sigma_x^2 = \Delta x^2 \frac{t}{\Delta t} \tag{3.28}$$

Replacing (3.27) in (3.28) it is found that

$$\Delta x^2 = 2D\Delta t \tag{3.29a}$$

or

$$\left(\frac{\Delta x}{\Delta t}\right)^2 = U^2 = \frac{2D}{\Delta t} \tag{3.29b}$$

where U can be interpreted as the velocity of the particles executing a random walk. Equation (3.29b) supplies a formal relation between the diffusion coefficient and the velocity of the randomly moving particles.

If the velocity of each particle is selected via a random number from a sample following the uniform distribution over a range of velocities from $-U_r$ to $+U_r$; it can easily be proved that the velocity U and the range U_r are related by

$$U_r^2 = 3U^2 \rightarrow U_r = \sqrt{\left(\frac{6D}{\Delta t}\right)} \tag{3.31}$$

The generation of that sample of random velocity values, subject to certain statistical constraints, is a case of Monte Carlo sampling, a procedure with wide importance in simulation theory and described briefly in the next section.

3.3.3 Monte Carlo sampling

Suppose that in a simulation experiment it is necessary to apply values of a stochastic variable whose distribution is known (either the exact theoretical probability density function or the experimental frequency histogram and their integrals). From the known probability law

$$P(X < x) = F(x) \tag{3.32}$$

it is possible to construct a sample of N values of the stochastic variable $X(x_1 \cdots x_N)$ via (3.32) and a series of N random numbers distributed uniformly over the interval 0–1 $(R_1 \cdots R_N)$. The set of N random numbers can be either machine generated or by an algorithm for generation of quasi-random numbers. One such algorithm is the Lehmer algorithm, described by the inductive formula

$$R_n = R_{n-1}a(\bmod m), \qquad n = 1 \cdots N \tag{3.33}$$

For a, m and R_1, it is found that $a = 8t \pm 3$ (where t is integer), $m = 2^b$ (b integer) and R_1 (integer prime). Their normalisation $(0 < R_i < 1)$ can be done in various ways. Most modern microcomputers can generate random numbers which can be called by RND or RAND statements.

After the production of a sequence of random numbers, the values of the stochastic variable are generated from a probability curve, equating the R_i to the probability of the required x_i (a procedure illustrated in Fig. 3.3)

$$P(X < x_i) = R_i \rightarrow x_i \tag{3.34}$$

After the production of the x_i values their distribution can be tested against the required distribution (via the χ^2 test, for example). It is obvious that uniformity of distribution of the R_i's and an increase in the size N of the sample ensure the success of the sampling experiment.

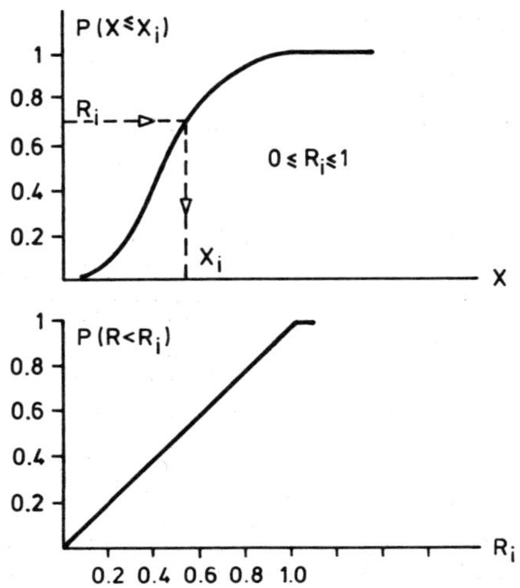

Fig. 3.3 Monte Carlo sampling from given cumulative distribution (uniform in the second case)

3.3.4 Solution algorithm and application

From the given analysis it can be deduced that simulation of the turbulent advective diffusion processes by the tracer technique contains the following discrete steps:

(1) The velocity field is determined by a set of values of velocity components at specific grid points.
(2) The time step Δt is selected.
(3) The range of random velocity $\pm U_r$ is computed from the diffusion coefficient D and the time step (3.31).
(4) The pollutant particles are positioned at the source. Two cases are distinguished. In the case of instantaneous discharge the whole number N (many thousands, for example) of the particles are positioned at the source with coordinates (X_0, Y_0). In the case of continuous discharge, the range $\Delta N/\Delta t$ is fixed and at each time step $\Delta t, \Delta N$ fresh particles are positioned at the (X_0, Y_0) location.
(5) Integration in time is executed and the new coordinates of each particle are computed. The motion of each particle is analysed into a deterministic and stochastic part

$$\Delta x_i^n = u(x_i^n, t^n)\Delta t \tag{3.35}$$

where $u(x_i^n, t^n)$ is the deterministic velocity at time t^n at the location

x_i^n of the ith particle, and

$$\Delta x_i^{n'} = u_i' \Delta t \tag{3.36}$$

where $u_i^{n'}$ is the random velocity at time t_n at the location x_i deriving from the described Monte Carlo sampling. The uniformity of the distribution of the random velocity variable results in the linear form of the probability function and direct analogy between R_i and $u_i^{n'}$ (see Fig. 3.3).

The new position of the particles is computed by the superposition of the deterministic and stochastic displacements

$$x_i^{n+1} = x_i^n + \Delta x_i^n + \Delta x_i^{n'} \tag{3.37}$$

In the case of a non-conservative pollutant the decomposition of the pollutant is realised by the selective disappearance of L number of particles. This number is analogous to the total number of particles in the field N, the time step Δt and the decay factor λ in (3.2).

$$L = \text{int}(N\lambda\Delta t) \tag{3.38}$$

In the case of a continuous discharge of a non-conservative pollutant these L points can be returned to the origin X_0, Y_0 (pollution source) so that the entrance rate equals the decomposition rate and the phenomenon evolves towards a steady state (stabilisation of the concentration values after a period of time). The organisation of the solution algorithm for a one-dimensional case in a computer program is given in Program 17 with an application which demonstrates the sensitivity of the solution to the D and Δt values.

The extension to two- or three-dimensional space is straightforward. The deterministic parts $\Delta x_i^n, \Delta y_i^n$ of the horizontal displacements of the ith particle are computed from the known horizontal velocity components u_i^n, v_i^n. The stochastic parts $\Delta' x_i^n, \Delta' y_i^n, \Delta' z_i^n$ are computed from the corresponding random velocity ranges (actually from the diffusion coefficients, Dx, Dy, Dz).

PROGRAM 17: TRACER TECHNIQUE FOR A 1-D ADVECTIVE DIFFUSION MODEL

```
   10REM TRACER TECHNIQUE FOR 1-D ADVECTION DIFFUS
ION MODEL
   20DIMX(1000),U(50),G(50)
   30READDT,DX,IM,X0,NM ,DIF
   40DATA...
   50FORI=1TOIM:READU(I):NEXTI
   60DATA...
   70FORN=1TONM:X(N)=X0:NEXTN
```

```
  80DU=SQR(6*DIF/DT)
  90T=0
 100T=T+DT
 110FORN=1TONM:I=INT(X(N)/DX)+1:UU=U(I)+(U(I+1)-U
(I))*(X(N)-(I-1)*DX)/DX
 120X(N)=X(N)+(UU+2*(RND(1)-.5)*DU)*DT:NEXTN
 130FORI=1TOIM:G(I)=0:NEXTI:FORN=1TONM:I=INT(X(N)
/DX)+1:G(I)=G(I)+1/NM:NEXTN
 140IF T/DT/10<>INT(T/DT/10) THEN GOTO 100
 150PRINTT:PRINT:FORI=1TOIM-1:PRINTG(I);:NEXTI:PR
INT
 160IFT/DT<NM THEN GOTO100
 170END
```

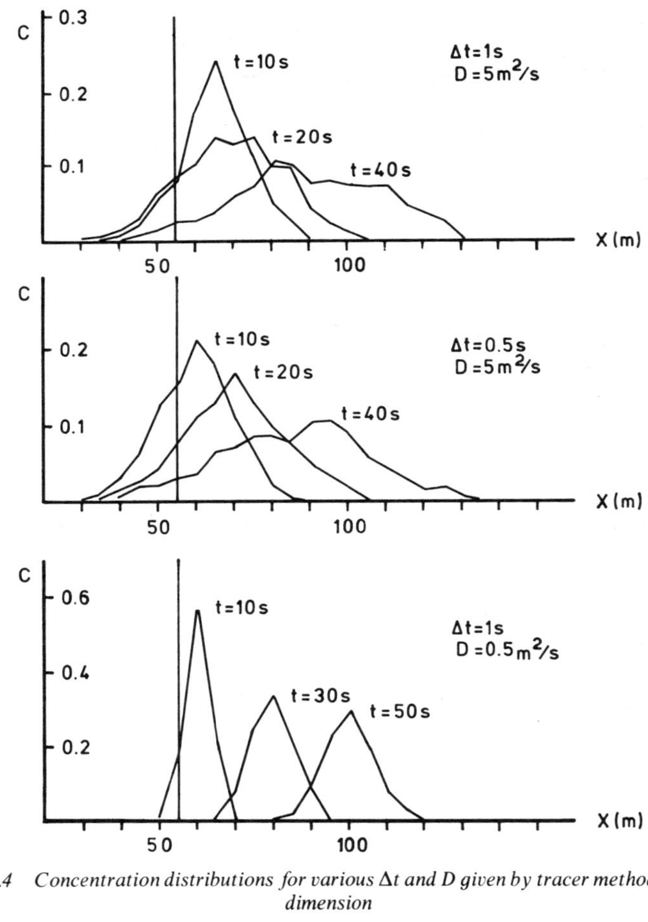

Fig. 3.4 Concentration distributions for various Δt and D given by tracer method in one dimension

Description of variables:

DT, DX = time step and space discretisation step
NM = number of pollutant's particles
IM = number of nodes for the discretisation of the 1-D flow domain
X0 = abscissa of the pollution source
NM = number of time steps in the experiment
DIF = diffusion coefficient
U(I) = fluid velocity at node $\# I$.

The printed results give the pollutant concentration referring to the space between two successive nodes. It is a relative concentration expressed as the ratio of the number of particles in that space to NM.

The application was done for uniform velocity, $u = 1\,\text{m/s}$ and $Dx = 5\,\text{m}$. The solution was repeated for two diffusion coefficient values, $D = 0.5\,\text{m}^2/\text{s}$ and $D = 5\,\text{m}^2/\text{s}$ and two Δt values, DT = 1 s and DT = 0.5 s. The number of simulation particles was NM = 500.

The graphical representation of the evolution of the pollutant concentration in Fig. 3.4 can be assessed by comparison with the analytical solution (Gaussian bell). The Péclet numbers for the two experiments were 1 and 10 respectively.

In conclusion it can be stated that:

(1) The method attains special importance for large Péclet numbers where advection dominates over diffusion.

(2) The accuracy of the results increases with the number of particles traced; several hundreds of particles would be an absolute minimum. As random numbers are introduced in the solution, the repetition of a simulation experiment does not lead to exactly the same results.

4

Mathematical models for sediment transport in coastal areas

4.1 PHYSICAL CONCEPTS—DEFINITIONS

The physical phenomena related to sediment transport in coastal areas and the resulting morphological changes of the coast (erosion, accretion) are major factors in the design, construction and operation of coastal structures. Research, directed toward the understanding of the physical processes involved in fluvial areas was initiated many decades ago. Empirical and mixed empirical–theoretical relations, such as the classical relations of DuBoys, Meyer–Meter, Engelund–Hansen, Einstein, Kalinske, etc. and, more recently, mathematical models, have been developed for the quantitative description of sediment motion.

The transfer of research experience into the coastal environment is a recent event, paralleled by the evolution of coastal engineering. The first observation was that the coexistence of waves and currents, in contrast to rivers which have a unidirectional current, made the direct transfer of river engineering experience inadequate. Thus, research into coastal sediment transport is continuously evolving.

The main difference, so far as sediment transport is concerned, between the river and the coast is that on the coast the waves act as a destabilisation, mobilisation and finally suspension factor for the sediment and a minimal current (even in the form of Stokes wave drift) is able to carry in the already activated sediment grains. However, in a river there is a certain threshold of bed shear stress and consequently of current velocity needed to initiate the motion.

This chapter is concerned with the presentation of the most basic but operational mathematical models in use today for the description of sediment motion in coastal areas.

For convenience of developing these mathematical models and not for physical reasons, various types of modes of sediment motion will be distinguished. Two types of sediment motion are usually assumed:

(1) Bed motion. This type of motion refers to grains that either roll or slip in contact with the bed or saltate with short jumps touching the bed from time to time. The quantity (mass or volume) of sediment that moves in this way forms the bed load. The width of the layer inside which this motion occurs is hardly distinguishable, but for dimensional reasons it is assumed to be a multiple of the grain diameter.
(2) Motion in suspension. The concentration of sediment in the carrying fluid (the concentration is defined similarly to that in the previous chapter as the load or volume of sediment in the unit volume of the mixture) is very high near the bed and decreases rapidly away from it. In part of the sediment, grains may exist that move in the fluid body without any contact with the bed. The forces that keep these grains in continuous suspension are hydrostatic buoyancy and hydrodynamic lift. The latter, in the case of turbulent flow, depends on the w' component (turbulent vertical fluctuations). The w' magnitude is related to the turbulence intensity and thus to the bed shear stress (τ_b). The force of gravity acting on the suspension results in a certain settling velocity w_f (related to the specific gravity, shape of the grains and density of the fluid). The suspended load is due to the sediment volume moving according to this mode of motion and depends on the ratio u_{b*}/w_f ($u_{b*} = \sqrt{(\tau_b/\rho)}$).

In the horizontal direction a distinction is made between the sediment motion in the surf zone and the remainder of the coastal zone.

(1) In the narrow coastal strip between the breaker line and the coastline the turbulence is very intense due to wave break. In this zone, large quantities of sediment are transported in the direction of the longshore current (littoral drift). The physical processes there are so complicated that research has been directed towards the integral computation of the total sediment load moving in that zone and its relation to the wave breaking characteristics.
(2) In the rest of the two-dimensional coastal domain where coastal structures also extend, non-negligible sediment transport due to the combined effects of waves and currents is taking place. It has been estimated that it is necessary to investigate the sediment transport process in depths up to at least 10 m (influenced by wind generated waves with period $T < 10$ s).

Mathematical models referring to the bed and suspended load, and to the littoral and greater coastal zones will be briefly presented in the following sections.

4.2 FORMULATION OF PARTICULAR MODELS FOR SEDIMENT TRANSPORT

4.2.1 The general sediment budget equation

The mathematical relation for predicting the evolution of a coastline or the water depth in a coastal area with considerable sediment transport derives from the principle of mass conservation for the moving grains. Defining the total volume discharge (including voids) of sediment along the x and y horizontal directions, by q_{sx}, q_{sy}, q_{bx}, q_{by}, where s refers to 'suspended' and b to 'bed load' ($[q] = \text{m}^3/\text{m/s}$) then, for a nearly horizontal bed (mild slope assumption), the time rate of change of the bed level ζ_b is given by

$$\frac{\partial \zeta_b}{\partial t} + \frac{\partial}{\partial x}(q_{sx} + q_{bx}) + \frac{\partial}{\partial y}(q_{sy} + q_{by}) = S \tag{4.1}$$

where S is a source or sink term. This term describes the rate of storage of sediment in suspension in a water column with dimensions $\mathrm{d}x, \mathrm{d}y, h$.

$$S = \frac{\partial}{\partial t}\int_{-h}^{0} c \, \mathrm{d}z \tag{4.2}$$

where $c(x, y, z, t)$ is the ratio of the suspended volume to the volume of the mixture. S is usually negligible. The relation between the volume of sediment discharge to the weight discharge, g_s, g_b, is

$$g_s = (1 - n)\gamma_s \cdot q_s \tag{4.3}$$

where n is the porosity, γ_s the grain specific weight (for example, for quartz sand, $\gamma_s = 2.6t/\text{m}^3$).

In Equation (4.1) the sum $q_s + q_b$ forms the total sediment load or more precisely, the total specific sediment discharge q_t.

$$q_t = q_s + q_b \quad (\text{m}^3/\text{m/s}) \tag{4.4}$$

Equation (4.1) is generally applicable to problems of bed formation in one or two dimensions.

4.2.2 Mathematical models for bed load

It was mentioned above that this part of the sediment load contains grains that roll, slip, or saltate over the bed. Its thickness is limited to a narrow zone above the bed. In the case of a smooth bed this width is assumed to average $(2\text{–}3)D_{50}$. Research into river engineering has established a direct correlation between the bed load (specific discharge) q_b and the bed shear τ_b due to the unidirectional movement

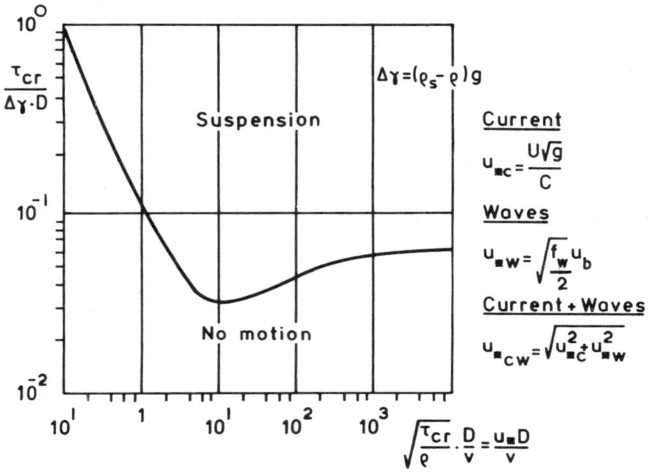

Fig. 4.1 Shield's diagram for initiation of sediment motion under various transport mechanisms

of the water. The motion of the grains starts when the value of τ_b is greater than the critical bed shear, τ_{cr}. The critical shear stress is computed from Shields diagram, valid and useful for almost half a century.

Recent research has shown that Shields diagram can be used even in the case of waves if the maximum friction velocity due to waves, given by Jonsson (see Fig. 4.1) is used for the friction velocity. For the application of this diagram, the mean grain diameter D_{50} and density ρ_s are necessary. The application of the critical shear stress concept in conjunction with information given by the higher order wave theory in shallow water (where the velocity and free surface oscillations are not symmetric with respect to the flow condition) results in a net movement of the grains during a wave period. The situation is illustrated in a simplistic way in Fig. 4.2.

This conceptual model, of course, is not valid in the case of a rippled bed where eddies with horizontal axes of rotation are generated over the rippled slopes during the passage of a progressive wave. This requires a different physical mechanism to explain the net transport of sediment. In all large scale mathematical models these local phenomena are not analysed and the parameters used tend to correlate sediment discharge to local bed shear conditions.

The algebraic relation between bed shear and bed load implies an assumption that inertial effects are negligible and that the grains respond without any hysteresis to the instantaneous hydrodynamic

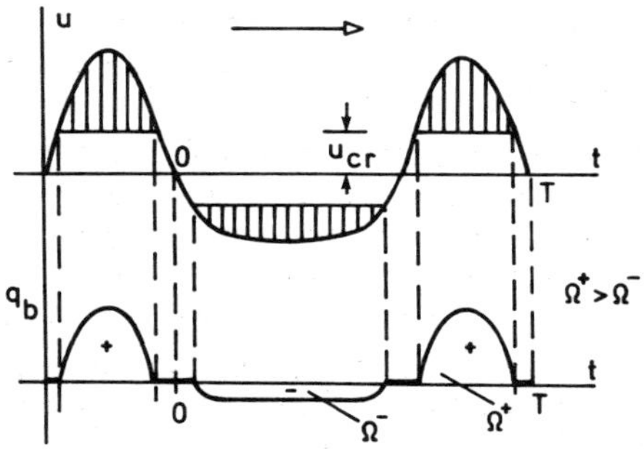

Fig. 4.2 Conceptual model for sediment transport by progressive waves

condition in the bed. This assumption is considered as realistic in most situations.

A historical review demonstrates that the DuBoys relation was one of the initial simple practical relations for the quantitative description of bed load,

$$q_b = x \cdot \tau_b(\tau_b - \tau_{cr}) \tag{4.5}$$

where the x coefficient is dimensional and its value is a function of the size, geometry and specific weight of the grains. The Meyer–Peter and Kalinske–Frijlink formulae came later and the latter applied today in Europe has the form:

$$q_b = 5D_{50}\frac{U}{C}\mu\sqrt{g}\exp\left[\frac{-0.27\Delta\rho C^2 D_{50}}{\mu U^2}\right] \tag{4.6}$$

where U is the mean current velocity, $\Delta\rho = (\rho_s - \rho_w)/\rho_w$ is the relative density of the grains, μ is a coefficient introducing the influence of the sand ripples if any, given by the relation

$$\mu = \left(\frac{C}{C'}\right)^{3/2} \tag{4.7}$$

where C is the Chézy friction coefficient for the rippled bed

$$C = 18\log(12h/k_s) \tag{4.8}$$

(h = depth, k_s = absolute bed roughness equal to half the height of the wave ripples) and

$$C' = 18\log(12h/D_{90}) \tag{4.9}$$

Equation (4.6) contains a product of two factors, the first called the transport factor and the second, of exponential form, called the stirring factor. Equation (4.6) gives the volumetric specific discharge of near-bed moving sediment in a current-dominated environment. In the case of currents and waves, this formula can be modified to take into account, in the stirring part, the wave effect. In this case, according to the analysis of Bijker the total bed shear stress (due to waves plus current) averaged during a wave cycle can be written as

$$\tau_{cw} = \tau_c\left[1 + \frac{1}{2}\left(\xi\frac{\hat{u}_b}{U}\right)^2\right] \tag{4.10}$$

where τ_c is the bed shear due to the current ($\tau_c = \rho g U^2/C^2$). The correction factor imposed on τ_c contains $\hat{u}_b$, the amplitude of wave velocity near the bed, (1st order theory may be used and thus for water of intermediate depth $\hat{u}_b = (\pi H/T)(\sinh kh)^{-1}$) and a dimensionless parameter ξ related to the bed roughness

$$\xi = \frac{c}{\sqrt{(2g)}}\sqrt{(f_w)} \tag{4.11}$$

where f_w is the bed friction coefficient for waves used in the Jonsson formula. Substitution of the expression (4.10) in the Kalinske–Frijlink equation leads to a modified bed load formula for waves plus current,

$$q_b = 5D_{50}\frac{U}{C}\sqrt{g}\exp\left[\frac{-0.27\Delta\rho D_{50}C^2}{\mu U^2\left(1 + \frac{1}{2}\left(\xi\frac{\hat{u}_b}{U}\right)^2\right)}\right] \tag{4.12}$$

It is not claimed that (4.12) has a universal validity.

Recently the Ackers and White formula has been given considerable attention and its validity investigated. This formula relates the total sediment transport to the sediment and current characteristics and has the form,

$$q_t = \frac{U}{1-n}D_{35}\left(\frac{U}{u_*}\right)^\alpha\frac{CD_{\mathrm{gr}}}{A^m}(F_c - A)^m \tag{4.13}$$

where D_{35}, CD_{gr} are measures of sediment grain, A is dimensionless parameter and F_c is the mobility factor,

$$F_c = \frac{U\left(\frac{u_*}{U}\right)^\alpha C^{\alpha-1}}{g^{\alpha/2}(\Delta\rho D_{35})^{1/2}} \tag{4.14}$$

The modified Kalinske–Frijlink formula will be used in the subsequent investigation of suspended sediment transport in the context of a general sediment transport formulation known as Bijker's method; practical, simple and of reasonable accuracy.

4.2.3 Mathematical model for suspended sediment load

Depending on the weight and shape characteristics of the sediment grains and the turbulence intensity of the carrying flow, sediment transport may occur without any contact with the bed but in the form of a suspension. This requires that the fall velocity is on average smaller than the vertical velocity turbulent fluctuations ($\sqrt{\bar{w}'^2}$). The dimensionless magnitude quantifying this criterion is the z_* factor defined ny

$$z_* = \frac{w_f}{\beta \kappa u u_*} \tag{4.15}$$

where β is the ratio ε/v of the sediment eddy diffusivity ε (symbolised as D in the previous chapter for pollutant diffusion) and the eddy viscosity coefficient v_v and κ the Von Karman constant ($\simeq 0.4$).

The fall velocity is described by the following relation

$$w_f = \frac{1}{18}\frac{(\gamma_s - 1)gD_{50}^2}{v}, \qquad D_{50} < 100\mu \tag{4.16a}$$

$$w_f = \frac{10v}{D_{50}}\left\{\left(1 + \frac{0.01(\gamma_s - 1)gD_{50}^3}{v^2}\right)^{0.5} - 1\right\}, \qquad 100\mu < D_{50} < 1000\mu \tag{4.16b}$$

$$w_f = 1.1\{(\gamma_s - 1)gD_{50}\}^{0.5}, \qquad D_{50} > 1000\mu \tag{4.16c}$$

where γ_s the specific weight of the grains, D_{50} the mean diameter and v the kinematic molecular viscosity coefficient of the water. Assuming that the suspended material behaves exactly as the dissolved matter in a turbulent advective diffusion situation ($\varepsilon \simeq D$) the field equation for the suspended sediment transport in a quasi-horizontal flow situation can be written

$$\frac{\partial c}{\partial t} + \frac{\partial (cu)}{\partial x} + \frac{\partial (cv)}{\partial y} - w_f\frac{\partial c}{\partial z} = \frac{\partial}{\partial x}\left(\varepsilon_h \frac{\partial c}{\partial x}\right) + \frac{\partial}{\partial y}\left(\varepsilon_h \frac{\partial c}{\partial y}\right) + \frac{\partial}{\partial z}\left(\varepsilon_v \frac{\partial c}{\partial z}\right) \tag{4.17}$$

The horizontal mass diffusion terms may be negligible compared with the vertical diffusion term as in the case of momentum transport in nearly horizontal geophysical flows. The magnitude of the vertical eddy diffusion coefficient is further analysed below. According to the definition of concentration, c defines the volume or submerged weight of sediment in unit volume of the mixture (water + grains). In the case of volume concentration, the suspended sediment load (specific discharge) is given by the integral over the depth,

$$q_{sx} = \int_{-h+a}^{0} uc\,dz, \qquad q_{sy} = \int_{-h+a}^{0} vc\,dz \tag{4.18}$$

The lower limit of integration is not on the bed but at a distance a equal to the thickness of the bed load transport zone. The mean concentration of sediment over the near bed zone is defined as

$$c_a = \frac{q_b}{a \cdot u_a} \tag{4.19a}$$

where u_a is the mean velocity of the grain moving as a bed load. For the mean bed concentration, c_a, Bijker suggests a more specific formula

$$c_a = \frac{q_b}{6.34\sqrt{\left(\frac{\tau_b}{\rho}\right)}k_s} \tag{4.19b}$$

where $\sqrt{(\tau_b/\rho)}$ is the bed friction velocity and k_s is half the height of the sand ripples. The concentration distribution over the depth and the artificial distinction between bed layer and suspension zone is shown schematically in Fig. 4.3.

On the basis of Equation (4.19) the bed boundary condition for the suspended sediment model is fixed.

At a distance a from the bed the rate of sediment deposition is given by the product $w_\varepsilon c$. The suspension rate is described by the product $\varepsilon_v\, \partial c/\partial z$

$$D = w_f c\Big|_{z=-h+a}, \qquad E = -\varepsilon_v \frac{\partial c}{\partial z}\Big|_{z=-h+a} \tag{4.20}$$

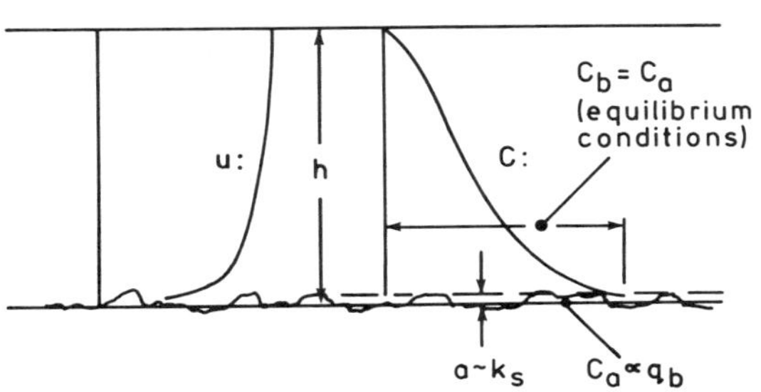

Fig. 4.3 Velocity distribution, bed and suspended load distributions in 1-D uniform flow case

The bed boundary condition has the general form

$$-\varepsilon_v \frac{\partial c}{\partial z}\bigg|_{-h+a} = w_f\left(c^\infty + Rc\bigg|_{-h+a}\right) \tag{4.21}$$

where c^∞ is the equilibrium concentration at that level (when $D \simeq E$). In the equilibrium case, $c^\infty = c_a$, where c_a is given by (4.19). The coefficient R describes the resuspension of settled sediment grains. For sandy material ($D > 100\mu$) that term may be neglected and (4.21) becomes

$$E = -\varepsilon_v \frac{\partial c}{\partial z}\bigg|_{-h+a} = w_f c^\infty = w_f c_a \tag{4.22}$$

Equation (4.22) gives the suspension rate near the bed. From the computation of $c|_{z=-h+a}$ the deposition rate D can be estimated. Their difference, $D - E$, gives the rate of bed level variation with time $(\partial\zeta_b/\partial t)$. On the free surface, as no sediment can cross that boundary, the surface boundary condition describes the equality of deposition and suspension.

$$w_f c\bigg|_{z=0} = -\varepsilon_v \frac{\partial c}{\partial z}\bigg|_{z=0} \tag{4.23}$$

On the lateral boundaries the suspended sediment distributions are usually assumed known. On the solid boundaries the zero normal flux condition is applied

$$\frac{\partial c}{\partial n} = 0 \tag{4.24}$$

On an open sea boundary the normal flux can be considered uniform across that boundary

$$\frac{\partial}{\partial n}\left(\frac{\partial c}{\partial n}\right) = 0 \tag{4.25}$$

The turbulence closure problem remains to be solved regarding the distribution of the eddy diffusion coefficient. This magnitude is related to the eddy viscosity coefficient and can be expressed as

$$\varepsilon_v = \varphi \cdot \beta \cdot \nu_v \tag{4.26}$$

where β is a dimensionless proportionality factor (it was found that $2 > \beta > 1$) describing the difference of the behaviour of fluid and solid particles in an eddy area and φ is also a dimensionless constant describing the influence of the high concentrations on the hydrodynamic state. For $O[c] < 0.1$ the influence of φ is negligible.

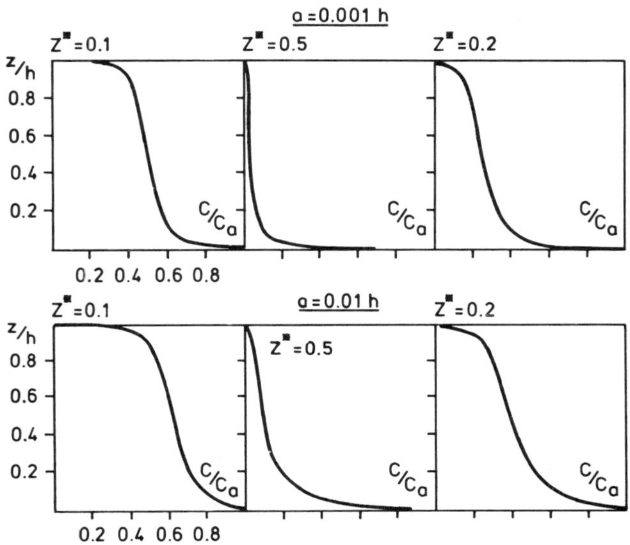

Fig. 4.4 Suspended sediment distributions for various z factor and q/h values*

For uniform free surface flow (4.26) can take the specific form

$$\varepsilon_v = \varphi \cdot \beta u_* h \cdot \kappa \frac{z}{h}\left(1 + \frac{z}{h}\right) \tag{4.27}$$

Recent experimental research suggests that for the upper half of the depth a constant ε value (equal to ε_{max}) is more realistic. For uniform flow and sediment equilibrium conditions (4.17) takes the form,

$$w_f c = -\varepsilon \frac{\partial c}{\partial z} \tag{4.28}$$

Its analytical solution for an $\varepsilon(z)$ distribution, given in (4.27), is

$$c(z) = c_a \left(\frac{h - z}{z} \frac{a}{h - a} \right)^{z_*} \tag{4.29}$$

where z_* is given in (4.15). The morphology of the $c(z)$ distribution for various a and z_* is shown in Fig. 4.4.

The $c(z)$ function is defined in the interval $z = 0$ to $z > -h + a$ (distance a from the bed). In the most general case of non-uniform flow and non-equilibrium state for the sediment, the numerical solution of (4.17) is indispensable for the computation of $c(x, y, z, t)$.

An approximation to this is computation of c locally, by means of (4.29), using the local hydrodynamic conditions.

Where both waves and currents occur, Equation (4.29) holds for uniform flow and sediment equilibrium, with modified z^* as $u_* = \sqrt{(\tau_c/\rho)}$ changes to $\sqrt{(\tau_{cw}/\rho)}$. Application of (4.10) in (4.29) leads to

$$c(z) = c_a\left(\frac{a}{h-a}\,\frac{h-z}{z}\right)^{z_*} \tag{4.31}$$

where

$$z_* = \frac{w_f\sqrt{\rho}}{\kappa\cdot\sqrt{\tau_c}\left\{\left(1+\frac{1}{2}\left(\xi\frac{\hat{u}_b}{U}\right)^2\right)\right\}^{1/2}}$$

For the computation of suspended sediment load q_s defined by (4.18), Equation (4.29) is applied in combination with the logarithmic $u(z)$ distribution valid for uniform steady flow,

$$u(z) = \frac{u_*}{k}\ln\left(\frac{h+z}{z_0}\right) \tag{4.32}$$

where $z_0 = k_s/33$ and the z ordinate is measured from the SWL upwards. The integral of (4.18), after the application of (4.29) and (4.32), can be written concisely as

$$q_s = Aq_b \tag{4.33}$$

where the A factor is a function of z_* and a/h. A nomograph for the estimate of A due to B. A. O'Connor is included in Fig. 4.5.

For z_* values, between 0.1 and 3, the suspended sediment load can be expressed approximately as a function of c_a, h and the depth mean current $U(=(1/h)\int u\,\mathrm{d}z)$ by the relation

$$q_s = \frac{\left[\left(\frac{a}{h}\right)^{z_*} - \left(\frac{a}{h}\right)^{1.2}\right]Uhc_a}{\left(1-\frac{a}{h}\right)^{z_*}(1.2-z_*)} \tag{4.34}$$

The sum $q_s + q_b$ is defined as the total load. There is a simple formula for the direct estimation of the total load, after Engelund and Hansen, usually leading to an overestimation of q_t,

$$q_t = q_s + q_b = 0.05U\frac{\tau_c^2 C}{\rho^2 g^{5/2}\Delta\rho^2 D_{50}} \tag{4.35}$$

This formula can be used in the case of waves and current, when τ_c is replaced by the corresponding τ_{cw}, given by (4.10).

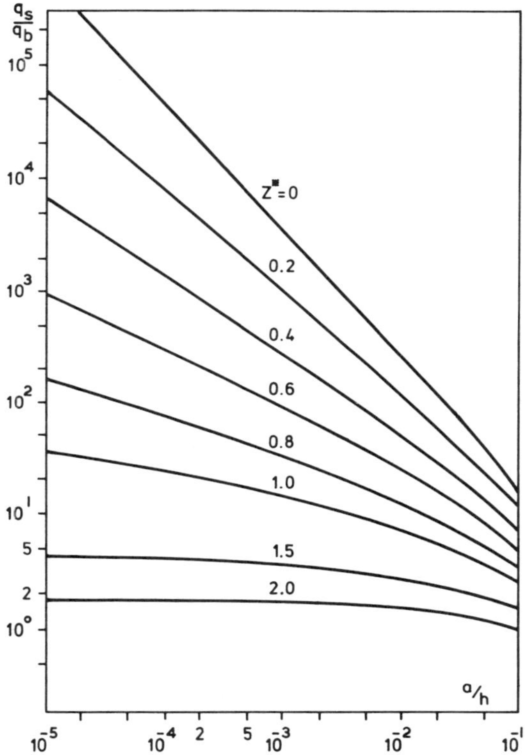

Fig. 4.5 Total load computation according to Einstein method

4.2.4 Sediment transport in the breaker zone

The oblique incidence of progressive waves on a sandy beach with slight bed slope results in the organisation of a discernible narrow zone extending from the first breaking line to the coastline (uprush limit), where the depth induced breaking of waves develops to its full extent. In this zone, of a width which may reach some hundreds of meters, the wave breaking phenomenon and the resulting turbulence destabilizes the bed grains and keeps a considerable amount of sediment in suspension. In the case of oblique breaking, the momentum component normal to the coast is counterbalanced by energy losses and positive surface gradients (wave set-up) while that parallel to the coast forms a longshore current developing inside the surf zone (Fig. 4.6).

The longshore current carries the destabilized and suspended

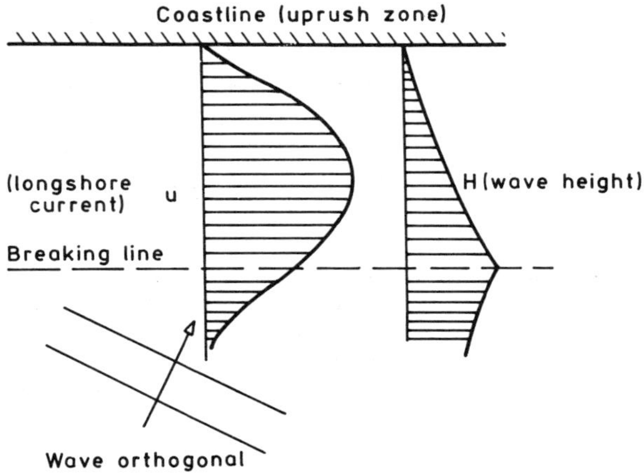

Fig. 4.6 Longshore current and wave height distribution across the surf zone

grains in its direction along the coast, forming the littoral transport. Sediment is carried by the breaking waves not only along the coast, but also normal to it. The result of this on–offshore transport is expressed in various ways:

(1) One or more longshore bars are formed on the bed at a distance from the coast due to the transport of sediment from the coast and from the open sea towards the bar. Generation of a bar stabilises the breaking line (causes further breaking over it) and protects the coast from further erosion.
(2) Beach material is sorted normal to the coast. Coarse grains remain near the coast while the fine grains suspended at mid-depth are carried offshore by the mid-depth currents (see Fig. 4.7).

Longshore sediment transport is a very important factor in the development of the coastal morphology in the case of coastal

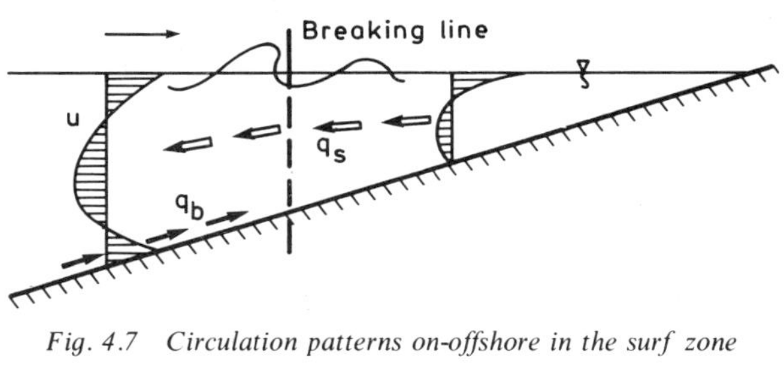

Fig. 4.7 Circulation patterns on-offshore in the surf zone

structures extending through the breaker zone. Considerable research effort has been devoted to its quantification. This research is based on field measurements and conservation of energy considerations. The established mathematical formulation contains a correlation of the sediment mass flux to the longshore energy flux. The relation has the general form:

$$S = AU \tag{4.36}$$

where S is the total volume discharge of sediment (bed + suspended load) inside the breaker zone, A is a correlation factor and U is the wave energy flux component along the coast. U is expressed as the product

$$U = Ec_g \tag{4.37}$$

where E is the longshore wave energy 'component' and c_g is the group velocity. From field and laboratory measurements, A is fixed at 0.014 (dimensionless) and Equation (4.36) takes the final form,

$$S = 0.014 H_0^2 c_0 k_{r_{br}} \cdot \sin \varphi_{br} \cdot \cos \varphi_{br} \tag{4.38}$$

where $[S] = m^3/s$, H_0 is a measure of wave height in the open sea, c_0 is the wave velocity in the open sea, $k_{r_{br}}$ is the refraction coefficient on the breaking line and φ_{br} is the angle between the coast and wave crests on the breaking line. For $[S] = m^3/\text{year}$, the coefficient $A = 0.44 \times 10^6$. For annual wave frequency f, (4.38) becomes

$$S = 0.44 \times 10^6 H_0^2 c_0 k_{r_{br}} \cdot \sin \varphi_b \cdot \cos \varphi_{br} \cdot f \tag{4.39}$$

For random (wind generated waves) following a Rayleigh distribution, H_{0s} is used in place of H_0. This is a CERC (Coastal Engineering Research Center, USA) formula widely used for longshore wave induced sediment transport. The criticism of (4.38) is focused on the following points:

(1) There is considerable scatter in the A values deduced from field measurements.
(2) Equation (4.38) does not consider the composition of the sand. It was derived from measurement on sandy beaches characterised by $175\mu < D_{50} < 1000\mu$. It does not consider the bed slope and the width of the surf zone.
(3) It gives no information on the distribution of S along the littoral transport zone.
(4) It does not apply to muddy and gravel beaches.

Every possible effort must be made to calibrate the CERC formula (determination of A) before its application to a specific coast.

4.3 EVOLUTION OF A DREDGED TRENCH. BED MORPHOLOGY MODELS

The prediction of the changing shape of the coastal bed in specific areas has, in the past, been accomplished exclusively by the use of hydraulic models with a movable bed. A major shortcoming was the scale effect and the inability to satisfy all dimensional analysis constraints. The recent development of a mathematical formulation for sediment transport, described above, has opened the way for the use of mathematical models for this purpose. The information required from a mathematical model of a sediment transport bed morphology is either the initial rate of change of bed morphology due to an imposed perturbation on a previous equilibrium state or the time dependent continuous evolution to a new equilibrium state.

The models themselves can be characterised according to the type of mathematical relations they use to three levels:

(1) Zero level models are those that use a simple formulation for the flow hydrodynamics deriving from the assumption of locally uniform flow and also, for the sediment transport, formulae giving local sediment load (bed + suspended) under equilibrium conditions as a function of local hydrodynamic action.
(2) First order models are those that use the same hydrodynamic formulation as above, but non-equilibrium transport formulae for the suspended sediment.
(3) Second order models are those that are required to solve the non-uniform flow equations and the non-equilibrium sediment transport equations for the computation of sediment load.

A critical presentation of these concepts follows for the problem of dredged cut evolution under cross-currents.

The approach to many coastal harbours constructed on sandy shallow coasts is achieved by means of a long port approach dredged channel (sometimes of the order of kilometers). This submarine channel is indicated by floats and lights and constitutes the only approach to the harbour basin. The coastal currents, in general oblique to the trench, lead to considerable deformation of its initial cross-section due to trapped sediments. If left unattended it would make the harbour unusable. The only solution is periodic dredging and maintenance of the approach channel which is a costly operation. However, the change in the channel section under the prevailing conditions of current and climate can be predicted by a mathematical model. When the sides of the channel slope steeply and the current is strong, the hydrodynamic field across the channel is highly non-uniform and the use of a 2nd order model is essential.

If the sides of the channel slope gently and the current is weak, then the flow across the channel may be considered uniform and a zero or first order bed morphology evolution model may be applied. The zero order model assumes the immediate adaptation of the sediment load to the local hydraulic situation and facilitates the mathematical computation, as it permits the use of local sediment transport formulae under equilibrium conditions but leads to the appearance of a singularity in the mathematical solution. As will be shown, the final form of the bed morphology model is that of a propagation of a non-linear wave of the bed elevation. Sooner or later an abrupt slope is formed on the bed; something that never happens in nature. The model behaves tolerably during the first stages of the bed formation, but later on, diffusive mechanisms which smooth the bed perturbation take place in nature. These mechanisms can be simulated numerically by considering the effect of the bed slope on the sediment discharge.

Consider the case of a current normal to the dredged channel (see Fig. 4.8). The complete bed morphology model should comprise:

(1) Hydrodynamic part

$$\frac{\partial u}{\partial t} + u\frac{\partial u}{\partial x} + w\frac{\partial u}{\partial z} = -g\frac{\partial \zeta}{\partial x} + \frac{\partial}{\partial z}\left(v\frac{\partial u}{\partial z}\right) \qquad (4.40)$$

$$\frac{\partial u}{\partial x} + \frac{\partial w}{\partial z} = 0 \qquad \text{or} \qquad \frac{\partial \zeta}{\partial t} + \frac{\partial}{\partial x}\int u\,\mathrm{d}z = 0 \qquad (4.41)$$

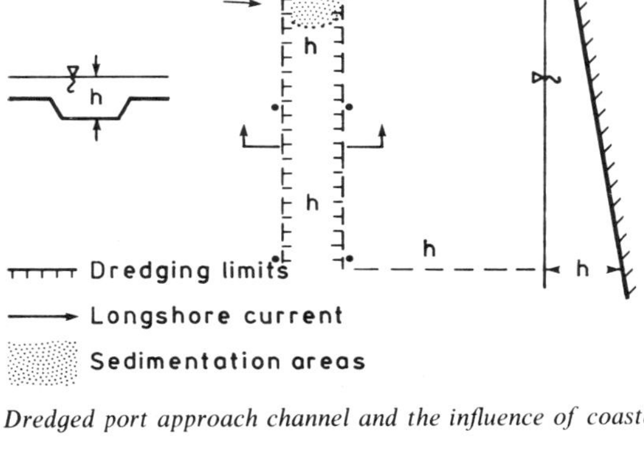

Fig. 4.8 Dredged port approach channel and the influence of coastal currents

The upstream boundary condition should describe the current profile away from the trench. Surface and bed boundary conditions are $\partial u/\partial z = 0$, $u = 0$, respectively.

(2) Sediment part

(i) For the bed load

$$q_b = f(h, \tau_b, D_{50} \cdots) \tag{4.42}$$

giving the bed reference concentration c_a.

(ii) For the suspended load

$$\frac{\partial c}{\partial t} + \frac{\partial (cu)}{\partial x} - w_f \frac{\partial c}{\partial z} = \frac{\partial}{\partial z}\left(\varepsilon \frac{\partial c}{\partial z}\right) \tag{4.43}$$

(iii) From the solution of (4.43) the computation of $c(z)$ and the subsequent computation of q_s follows:

$$q_s = \int_{-h+a}^{0} cu \, \mathrm{d}z \tag{4.44}$$

or computation of the difference

$$D - E = w_f \cdot c \Big|_{-h+a} - \varepsilon \frac{\partial c}{\partial z}\Big|_{-h+a} \tag{4.45}$$

(iv) Substitution in the sediment conservation formula (4.1)

$$\frac{\partial \zeta_b}{\partial t} + \frac{\partial}{\partial x}(q_s + q_b) = \frac{\partial \zeta_b}{\partial t} + \frac{\partial q_b}{\partial x} - D + E = 0 \tag{4.46}$$

The solution of the system of PDE's forming the 2nd order model is beyond the introductory scope of this book.

Let us assume here the extremely simplified situation of uniform flow with a velocity profile defined by the current specific discharge q and the local depth $h(x)$. The bed shear is computed by

$$\tau_b = \rho f U^2 = \rho f \left(\frac{q}{h}\right)^2 \tag{4.47}$$

where f a constant bed friction coefficient and q the constant discharge. For logarithmic velocity distribution, f is given by

$$f = \kappa \Big/ \left(\ln \frac{h}{z_0} - 1\right)^2 \tag{4.48}$$

For the sediment load the classical DuBoys formula is adopted. This is done for illustration purposes and it is obvious that the Engelund–Hansen total load formula would show the same behaviour. The substitution of Equation (4.47) into (4.6) gives:

$$q_b = x\tau_b(\tau_b - \tau_{cr}) = \frac{x\rho^2 f^2 q^4}{h^4} - \frac{x\rho f q^2 \tau_{cr}}{h^2} \tag{4.49}$$

The substitution of q_b from (4.49) to the sediment conservation equation (4.46) in the form

$$\frac{\partial h}{\partial t} - \frac{\partial q_b}{\partial x} = 0 \tag{4.50}$$

results in a PDE in the unknown $h(x, t)$ function

$$\frac{\partial h}{\partial t} - \frac{\partial h}{\partial x}\left(\frac{4x\rho^2 f^2 q^4}{h^5} - \frac{2x\rho f q^2 \tau_{cr}}{h^3}\right) = 0 \tag{4.51}$$

This is a first order non-linear hyperbolic equation, predicting that the h 'wave' is propagating in the direction of flow as a non-linear perturbation with velocity c, analogous to

$$c \propto \left(\frac{c_1}{h^5} + \frac{c_2}{h^3}\right) \tag{4.52}$$

For example, for $D_{50} = 0.1\,\text{mm}$, $\tau_{bc} = 0.07\,\text{kp/m}^2$, $x = 0.0021$, $f = 0.01$, $q = 0.5\,\text{m}^2/\text{s}$ and $h = 1\,\text{m}$, it is found that $c = 4 \times 10^{-4}\,\text{m/s}$. The numerical solution for this simple model of bed morphology follows.

The equations to be successively and repeatedly solved are rewritten:

$$\left.\begin{aligned} &U = q/h \\ &\tau_b = \rho f U^2 \\ &q_b = x\tau_b(\tau_b - \tau_{cr}) \\ &\frac{\partial h}{\partial t} - \frac{\partial q_b}{\partial x} = 0 \end{aligned}\right\} \text{(or any other local transport formula)} \tag{4.53}$$

The numerical solution of the model is performed by finite differences on a non-staggered grid. In the case of unidirectional flow the backward difference explicit scheme dominated by the stability criterion $c\Delta t/\Delta x < 1$ is used. Its numerical diffusion decreases with the increase of Courant number. For space and time indices i ($x_i = i\Delta x$) and n ($t_n = n\Delta t$), respectively, the model (4.53) is written

$$\left.\begin{aligned} &U_i^n = q/h_i^n \\ &\tau_{b_i}^n = \rho f (U_i^n)^2 \\ &q_{b_i}^n = x\tau_{b_i}^n(\tau_{b_i}^n \tau_{cr}) \\ &h_i^{n+1} = h_i^n + \frac{\Delta t}{\Delta x}(q_{b_i}^n - q_{b_{i-1}}^n) \end{aligned}\right\} \tag{4.54}$$

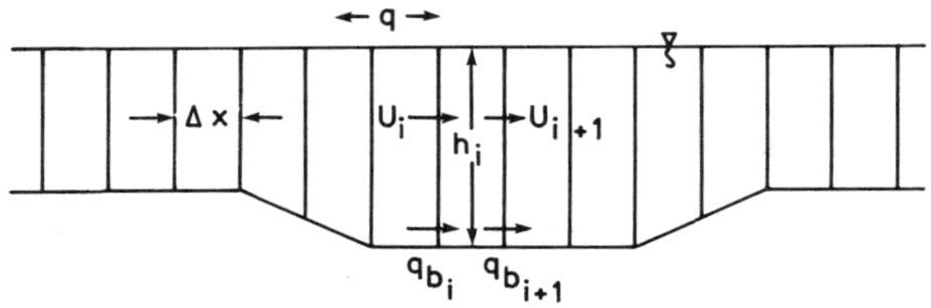

Fig. 4.9 Flow domain discretisation for 1-D bed morphology model

In the case of oscillating flow (tidal current over the dredged cut) the above scheme would not be consistent and the finite differences should follow the flow direction. Instead, a centered finite difference on a staggered grid (see Fig. 4.9) with controllable numerical diffusion gives smooth results. A Lax type difference for the time derivative can be used. The model (4.53) is written:

$$\left.\begin{aligned} U_i^n &= 2q/(h_i^n + h_{i-1}^n) \\ \tau_{b_i}^n &= \rho f (U_i^n)^2 \\ q_{b_i}^n &= x\tau_{b_i}^n(\tau_{b_i}^n - \tau_{cr}) \\ h_i^{n+1} &= h_i^n\theta + \frac{1-\theta}{2}(h_{i+1}^n + h_{i-1}^n) + \frac{\Delta t}{\Delta x}(q_{b_{i+1}}^n - q_{b_i}^n) \end{aligned}\right\} \quad (4.55)$$

The solution algorithm, Program 18, in BASIC is listed below.

PROGRAM 18: SEDIMENTATION OF DREDGED CUT, UNIDIRECTIONAL FLOW

```
    5REM SEDIMENTATION OF DREDGED CUT UNIDIRECTION
AL FLOW
   10DIMH(31),HN(31),QS(31)
   20READDX,DT,IM,NM,Q,CC,FR,TC,LX
   40DATA...
   50FORI=1TOIM:READH(I):NEXTI
   60DATA...
   70N=0:TT=0
   80N=N+1:TT=TT+DT
  100FORI=1TOIM:U=Q/H(I):T=FR*U*ABS(U)*100:IFABS(T
)>TC THENQS(I)=CC*T*ABS(T-TC) ELSE QS(I)=0
  110NEXTI
  120FORI=2TOIM:HN(I)=H(I)*LX+(1-LX)/2*(H(I+1)+H(I
-1))/2+DT/DX*(QS(I)-QS(I-1)):NEXTI:HN(1)=HN(2)
  130FORI=0TOIM:H(I)=HN(I):NEXTI
  140IFN/100<>INT(N/100) THEN GOTO80
  150PRINTN:PRINT:FORI=1TOIM:PRINTH(I);:NEXTI:PRIN
T
  160IFN<NM THEN GOTO 80
  170END
```

Description of main variables:

DX, DT = space and time discretisation steps
IM = number of cross-sections (U computation points)
NM = number of time steps
Q = amplitude of specific water discharge (m^2/s)
CC = constant X in duBoys formula ($m^6/kp^2\,s$)
FR = dimensionless friction coefficient
TC = critical bed shear stress (kp/m^2)
LX = weighting factor in Lax time difference.

The application refers to the deformation of a dredged trench under the influence of a constant current. The initial form and the discretisation of the trench is given in Fig. 4.10(a). The data for this application are, DX = 1 m, DT = 10 s, IM = 30, NM = 500, Q = 1 m^2/s, CC = 0.0021, FR = 0.01, TC = 0.07, LX = 0.1.

The depth formation is given in Fig. 4.10(b). The sensitivity of these results to horizontal diffusion entailed by the finite difference scheme is obvious. No trail effects are visible.

4.4 EVOLUTION OF COASTLINE DUE TO LITTORAL TRANSPORT

It was mentioned in the introductory paragraph to this chapter that a considerable percentage of the sediment transported in a coastal area moves in the form of a submarine river inside the strip of the surf zone.

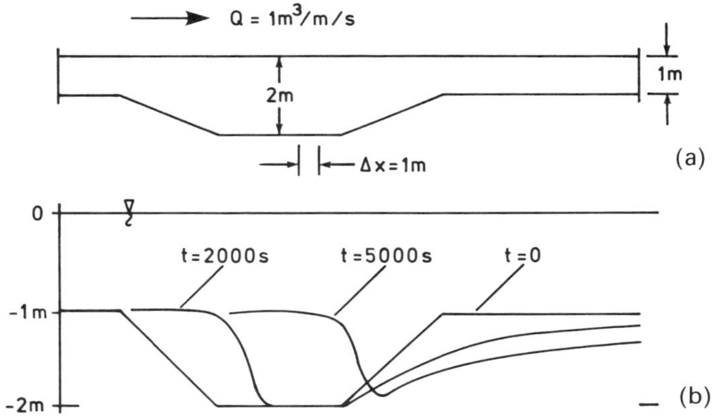

Fig. 4.10 Local transport bed morphology model results for dredged cut under cross current

The intervention of a coastal structure in the wave breaking process disturbs the pre-existing sedimentologic equilibrium. Either in the form of a mole or a detached breakwater, or a pair of jetties protecting inlets, the structure traps sediment and disturbs the flow of sediments along the coast. The long-term effect is the erosion or accretion of the coastline. The impact of coastal structures on the coastline is of the utmost concern where the coast has been developed for tourist or industrial purposes. It is essential to incorporate in the design a forecast of how the coastline will evolve in the neighbourhood of the structure. That neighbourhood, surprisingly, may extend over several miles on either side of the structure.

The most established mathematical relation for the littoral sediment transport due to oblique wave breaking is the CERC formula (4.38), giving the total load as a function of the incident significant wave height, breaking angle and breaking wave characteristics. This relation, in the case of depth contours nearly parallel to the coastline, can be expressed as a function of the incidence angle φ_0 and not the breaking angle φ_{br}. Equation (4.38) becomes,

$$S = 0.014 H_0^2 c_{br} \sin \varphi_0 \cos \varphi_0 \tag{4.56}$$

This equation reveals the functional relation $S(\varphi_0)$. This relation is not simple as c_{br} is also implicitly connected to φ_0. Two simplifying realistic assumptions may be made to that end:

(1) assume a breaking criterion $H/h_{br} = 0.8$
(2) assume, according to the 1st order shallow-water wave theory, $c = \sqrt{\{g(h + H/2)\}}$.

These two assumptions give for c_{br}

$$c_{br} \approx 4 H_0^{1/2} \tag{4.57}$$

The substitution of (4.57) in (4.56) shows that $S(\varphi_0)$ varies approximately as $\sin 2\varphi_0$ with φ_0. The classical mathematical model for predicting the evolution of a coastline due to longitudinal gradients of $S(x)$ is known as the Penlard–Considere theory (single line theory for a straight coast and unidirectional waves). That theory will be extended for any coastline configuration, in a numerically integrable form and a solution algorithm will be presented. The physical situation and the notation to be used are illustrated in Fig. 4.11.

The first definition concerns the total depth h where the influence of waves can reach in destabilising the coastal sediments. This depth can be estimated from the wave characteristics (H, T) and the sediment characteristics (D_{50}, ρ_s). A simple realistic procedure for the

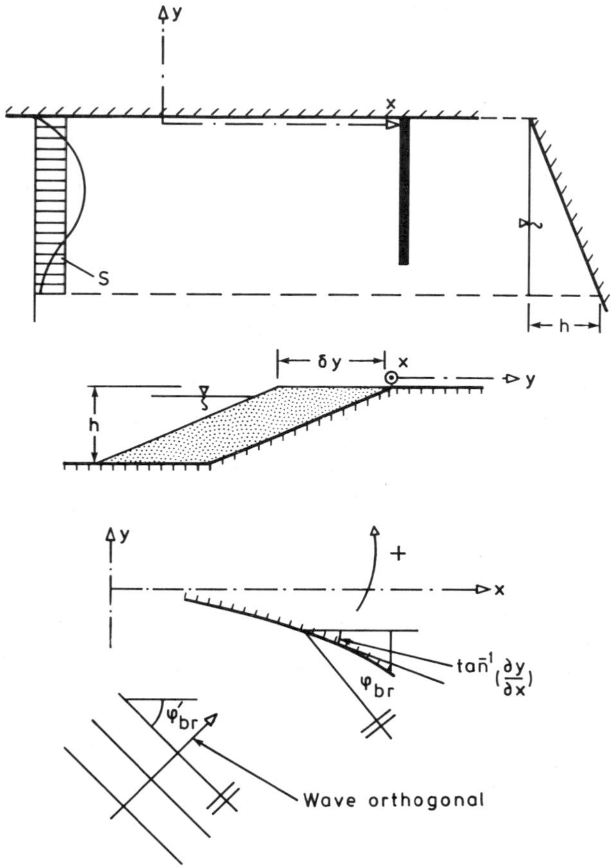

Fig. 4.11 Basic notation for generalised Penlard-Considere model

estimation of h comprises the following steps:

(1) Computation of the equivalent wave induced bed shear, according to Jonsson

$$\frac{\tau_w}{\rho} = \frac{f_w}{12\rho} u_{b_{\max}}^2 \tag{4.58}$$

(2) Application of Shields diagram using the value of bed shear from (1) for the computation of the critical value $\tau_{b_{cr}}$. Successive approximations for the computation of the h value for which the bed shear is critical.

The single line theory assumes that the bed deformation moves parallel to its original position from the coastline to a depth equal to h. If $y(x, t)$ is the coastline ordinate with respect to the x axis (in the general direction of the coast) then the sediment continuity equation written in terms of the volume discharge of sediment along the coast gives the temporal evolution of y.

$$\frac{\partial y}{\partial t} = \frac{1}{h}\frac{\partial S}{\partial x} \tag{4.59}$$

This equation assumes that the $y(x)$ curve has a gentle slope.

The model is completed with the dynamic equation supplying $S(x, t)$. The CERC formula (4.38) can be used in conjunction with (4.59). The wave breaking angle with respect to the coastline is expressed as the signed sum of the wave breaking angle with respect to the x axis (a function of x) and the angle that the coastline forms with the x axis (a function of x, t).

$$\varphi_{br} = -\varphi'_{br} + \tan^{-1}\frac{\partial y}{\partial x} \tag{4.60a}$$

Instead of using (4.38) in its row form, a differential equation for S can be formed based on (4.56). From a geometric relation similar to (4.60), but written in terms of the incidence angle φ_0,

$$\varphi_0 = -\varphi'_0 + \tan^{-1}\frac{\partial y}{\partial x} \tag{4.60b}$$

On application of the chain rule of differentiation to (4.56) it is found that

$$\frac{\partial S}{\partial t} = \frac{\partial S}{\partial \varphi_0}\cdot\frac{\partial \varphi_0}{\partial t} = \frac{\partial S}{\partial \varphi_0}\left(-\frac{\partial \varphi'_0}{\partial t} + \frac{\partial^2 y}{\partial t\,\partial x}\left(1 + \left(\frac{\partial y}{\partial x}\right)^2\right)^{-1}\right)$$

which for φ'_0 independent of t and substituting $\partial y/\partial t$ from (4.59) becomes

$$\frac{\partial S}{\partial t} = \frac{\partial S}{\partial \varphi_0}\left\{1 + \left(\frac{\partial y}{\partial x}\right)^2\right\}^{-1}\frac{1}{h}\frac{\partial^2 S}{\partial x^2} \tag{4.61}$$

Equation (4.61) is a parabolic equation in $S(x, t)$. The equivalent diffusion coefficient is

$$\sigma = \frac{\partial S}{\partial \varphi_0}\left(1 + \left(\frac{\partial y}{\partial x}\right)^2\right)^{-1}\frac{1}{h} \tag{4.62}$$

If $\partial S/\partial \varphi_0$ is determined either exactly or approximately, (Equations (4.56), (4.57)), the mathematical model consisting of (4.59), (4.60) and

(4.61) can be solved numerically. The boundary conditions completing the model are of three types:

(1) Far upstream or downstream of the coastal structure perturbing the coastline, S may be assumed uniform and $y = \text{constant}$ $(\partial y/\partial t = 0)$.
(2) At a coastal structure extending across the breaker line, $S = 0$ initially.
(3) In certain locations some sediment may bypass a coastal structure (as in the case of a dredged trench or a mole accreted on one side); there a functional relation $S(y)$ has to be provided.

The initial condition of the model consists of the previously known values of $y(x)$ and $S(x)$.

The integration in time is performed numerically by the computation of $y(x_i, t_n)$, $S(x_i, t_n)$ values at characteristic locations indicated by the i, n indices in x, t space. The goal of this model is to provide a prediction of the long term (annual or seasonal) variation, y, in the coastline under the existing wave climate. The forcing factors for the y variation are wave events characterised by specific values H_{0s}, period T, angle of incidence φ_0' and duration. For the activation of the model a time series of weather (waves) conditions must be known. This may be a recorded actual time series or artificially produced from processed wave or wind data (intensity-frequency-direction). The artificial time series is produced by Monte Carlo sampling according to the procedure described in Section 3.3.3.

The discretisation of the coastal domain is done in 1-D elements. A staggered grid is used in the present algorithm. The sediment discharge is computed at the cross-sections and the coastline ordinate refers to a reach between two cross-sections.

An explicit finite difference scheme is used for the sediment continuity equation.

$$y_i^{n+1} = y_i^n \theta + \frac{1-\theta}{2}(y_{i+1}^n + y_{i-1}^n) + \frac{(S_{i+1}^n - S_i^n)\Delta t}{\Delta x h} \tag{4.63}$$

The dynamic equation for the computation of the sediment discharge can be either the CERC formula or its modified form (parabolic PDE in S (4.61)). In the first case, an explicit finite difference approximation is

$$S_i^{n'} = 0.014 H_0^2 c_0^2 k_{r_{br_i}}^2 \cos \varphi_{br_i} \sin \varphi_{br_i} \tag{4.64}$$

with the φ_{br_i} given by the geometric relation (4.60a)

$$\varphi_{br_i} = -\varphi_{br_i}' + \tan^{-1}\left(\frac{y_i^n - y_{i-1}^n}{\Delta x}\right) \tag{4.65}$$

In the second case, when the PDE in S is used, an implicit scheme is recommended in order to overcome the stability limit on Δt ($\Delta t < \Delta x^2/(2\sigma)$). Equation (4.61) is written in finite difference form.

$$S_i^{n+1} = S_i^n + \frac{\Delta t \sigma_i^n}{\Delta x^2}(S_{i+1}^{n+1} - 2S_i^{n+1} + S_{i-1}^{n+1}) \tag{4.66}$$

where according to (4.62)

$$\sigma_i^n = 0.014 H_0^{2.5} \frac{1}{h} \cos(2\varphi_{0i}^n)\left\{1 + \left(\frac{y_i^n - y_{i-1}^n}{\Delta x}\right)^2\right\}^{-1} \tag{4.67}$$

where φ_{0i}^n is given, according to (4.65)

$$\varphi_{0i}^n = -\varphi_{0i}^{n'} + \tan^{-1}\left(\frac{y_i^n - y_{i-1}^n}{\Delta x}\right) \tag{4.68}$$

The BASIC code presented in Program 19 corresponds to a model consisting of Equations (4.63), (4.64) and (4.65) and refers to the case of a straight coastline with a mole in the middle interrupting the sediment flow. The evolution of accretion and erosion of the coastline on the two sides of the mole is followed for some time under constant wave conditions defined by H_{0s}, φ_0 and given h (depth of influence). The generalisation for varying wave conditions and different coastline-coastal structures configuration is left to the reader.

PROGRAM 19: CONSIDERE GENERALISED MODEL

```
   10REM CONSIDERE GENERALISED MODEL
   20DIMS(31),A(31),AB(31),HB(31),Y(31),YN(31),K(3
1)
   30READDT,DX,HORMS,IM,NM,H
   40DATA...
   50FORI=1TOIM:READY(I):NEXTI
   60DATA...
   70FORI=1TOIM:READAB(I):NEXTI
   80DATA...
   90FORI=1TOIM:READK(I):NEXTI
  100DATA...
  105FORI=1TOIM:HB(I)=HORMS*K(I):NEXTI
  110N=0:T=0
  120N=N+1:T=T+DT
  130FORI=2TOIM-1:A(I)=-RAD(AB(I))+ATN((Y(I)-Y(I-1
))/DX):S(I)=.083*HB(I)^2.5*SIN(2*A(I)):NEXTI
  140S(16)=0:S(1)=S(2):S(IM)=S(IM-1)
  150FORI=1TOIM-1:YN(I)=Y(I)+DT/DX/H*(S(I+1)-S(I))
:NEXTI:FORI=1TOIM:Y(I)=YN(I):NEXTI
  160IFN/100<>INT(N/100) THEN GOTO120
```

```
170PRINTN:FORI=1TOIM:PRINTY(I);:NEXTI:PRINT
180IFN<NM THEN GOTO120
190END
```

Description of main variables:

DT, DX = discretisation steps (m, s)
HORMS = open sea root mean square wave height (m)
Ab(I) = wave breaking angle with respect to the *Ox* axis (degrees)
Y(I) = coastline ordinates (changing in time) (m)
MN = number of time steps
IM = number of discretisation nodes
H = influence depth (m)

The application field and the resulting coastline evolution under continuous wave breaking is illustrated in Fig. 4.12.

The basic data is: DT = 50 s, DX = 10 m, HS = 1 m, NM = 1000, IM = 31, H = 5 m. The wave breaking angle (with respect to *x* axis) array is included in Fig. 4.12.

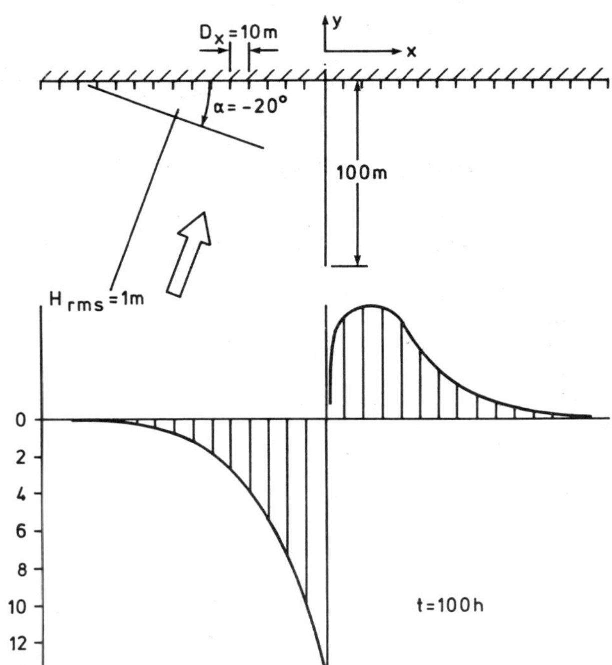

Fig. 4.12 Coastline evolution for longshore transport interrupted by mole

Concluding with the single line, generalised Considere model, the shortcomings limiting its applicability are stressed once more:

(1) As it makes use of CERC formula, the model is confined within the limits of the formula validity, i.e. small incidence angles φ_0, non-universality of the A factor and no consideration of the bed slope and beach material.
(2) The boundary conditions on the coastal structures and the lateral limits have to be fixed with the maximum accuracy.
(3) The application for a series of wave conditions simulating the local coastal climate have to be based on a realistic time series.
(4) A controllable numerical diffusion has to be used with utmost care to avoid spurious oscillations on the coastline that may result from interchanging wave incidence directions.

4.5 BED MORPHOLOGY EVOLUTION MODEL IN GREATER COASTAL DOMAINS

Outside the breaker zone, where the littoral longshore transport is the main factor of sediment transport and evolution of the coastline, there is a two-dimensional coastal domain where sediment is also transported by currents and waves. This domain extends to a depth where the influence of waves and currents becomes negligible. In areas where there are no strong tidal currents this depth of wave influence on the sediments is the open sea boundary of the sediment transport domain.

The erection of a coastal structure in a two-dimensional coastal domain is an intervention in a pre-existing hydrodynamic and sedimentation status quo. In that context it is necessary to predict the impact that the engineering work may have on the bed morphology so that an optimal layout of the work is selected (the resulting bed accretion or erosion not damaging to the operation of the construction work itself).

Bed morphology models of the type mentioned in Section 4.3 which give the evolution of the bed until a new equilibrium state is reached, are needed. The present state of mathematical modelling does not make this prediction feasible. The scope is limited to a prediction of the rate of change of bed level during a climatological year under a full typical cycle of wave conditions.

The hydrodynamic events during a whole year have to be known (measured or hindcasted). With known sequence of U, V, H, T fields and known annual frequencies the sediment continuity equation in two horizontal dimensions (Equations (4.1)) gives the rate of change

$\partial\zeta_b/\partial t$ along the coastal domain. The total sediment loads q_{tx}, q_{ty} resulting from the combined action of waves and currents in that type of operational model are given either by the Bijker method (Equations (4.14) and (4.33)) or a total load formula (for example, Engelund and Hansen (4.35)).

The application of this sediment continuity equation gives the areas that tend to be eroded and those that tend to be accreted. For quite large Δt in the $\partial\zeta_b/\partial t$, the bed variation may be so large that the feedback to the hydrodynamic conditions has to be considered. With the interchanging computation of the U, V, H and $h_{\text{old}} + \Delta\zeta_b = h_{\text{new}}$ fields, the model may dvance in time to reach theoretically the new bed equilibrium state. This is not the case in reality, as was mentioned above, due to the limitations of the model and the errors accumulated in its solution.

The numerical solution of the mathematical model for the macroscopic evolution of the bed morphology, due to sediment transport, in a two-dimensional domain, is focussed on the solution of the sediment continuity equation. It can be accomplished by an explicit finite difference scheme on a staggered grid. The sediment specific total discharges, q_{tx} and q_{ty}, are computed on the mesh sides and the bed elevation at the mesh center.

Assuming the U, V values to be known on the mesh sides and the h, H values at the mesh centers, the synthesis of the computer program is straightforward. The following BASIC program can easily become a subroutine for any of the wind and wave generated circulation codes in Chapter 2. It makes use of the Engelund–Hansen total load formula (a local transport formula).

PROGRAM 20: 2-D BED MORPHOLOGY VARIATION RATE MODEL

```
10REM 2-D BED MORPHOLOGY VARIATION RATE MODEL
20DIMU(20,20),V(20,20),DZB(20,20),H(20,20),HW(2
0,20),QTX(20,20),QTY(20,20)
30READ IM,JM,D50,DT,FW,C,DR
40DATA...
50FORJ=1TOJM:FORI=1TOIM:READU(I,J),V(I,J),H(I,J
),HW(I,J):NEXTI:NEXTJ
60DATA...
65FORI=1TOIM-1:READDX(I):NEXTI
66DATA...
67FORJ=1TOJM-1:READDY(J):NEXTJ
68DATA...
70KSI=C*SQR(FW/2/9.81)
80FORI=1TOIM:FORJ=1TOJM:IFU(I,J)=0 THEN GOTO160
```

```
90UB=(HW(I,J)+HW(I-1,J))/2*SQR(9.81*2/(H(I,J)+H
(I-1,J)))
95VV=(V(I,J)+V(I-1,J)+V(I,J+1)+V(I-1,J+1))/4
100TCWX=U(I,J)*SQR(U(I,J)^2+VV^2)/C^2*9.81*(1+.5
*(KSI*UB/U(I,J))^2)
110QTX(I,J)=.05*U(I,J)*TCWX^2*C/9.81^2.5/DR^2/D5
0
120IFV(I,J)=0 THEN GOTO160
130VB=(HW(I,J)+HW(I,J-1))/2*SQR(9.81*2/(H(I,J)+H
(I,J-1)))
135UU=(U(I,J)+U(I+1,J)+U(I,J-1)+U(I+1,J-1))/4
140TCWY=V(I,J)*SQR(V(I,J)^2+UU^2)/C^2*9.81*(1+.5
*(KSI*VB/V(I,J))^2)
150QTY(I,J)=.05*V(I,J)*TCWY^2*C/9.81^2.5/DR^2/D5
0
160NEXTI:NEXTJ
170FORI=1TOIM-1:FORJ=1TOJM-1
180DZB(I,J)=-DT/DX(I)*(QTX(I+1,J)-QTX(I,J))-DT/D
Y(J)*(QTY(I,J+1)-QTY(I,J)):NEXTJ:NEXTI
190FORJ=JM-1 TO 1 STEP-1:FORI=1TOIM-1:PRINTDZB(I
,J);:NEXTI:PRINT:NEXTJ
200END
```

Description of main variables:

DT, DX = discretisation steps along time and space dimensions
D50 = the mean grain diameter
FW = bed friction coefficient due to waves

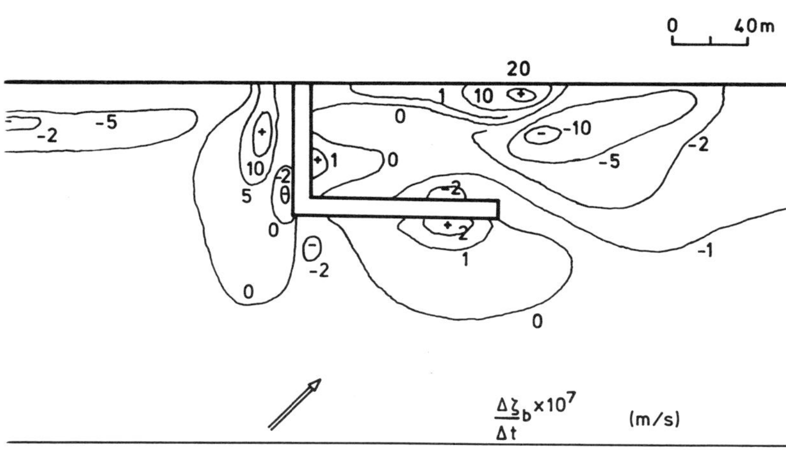

Fig. 4.13 2-D local transport bed morphology rate model. Current condition given in Fig. 2.19

C = Chézy coefficient
DR = density difference $(\rho_s - \rho_w)/\rho_w$
IM, JM = maximum values of I, J indices along x, y directions, respectively
H = water depth values refering to mesh centers
U, V = velocity components values (x, y) refering to mesh sides
HW = wave height values refering to mesh centers
DZB = rate of bed elevation variation values refering to mesh centers.

The application was done for the computation of the rate of accretion and erosion in the neighborhood of the mole of Section 2.4. The wave induced circulation is known for that case. The data in that example were, DR = 150, D50 = 0.0005 m, C = 45 $m^{\frac{1}{2}}/s$ and FW = 0.1.

Figure 4.13 gives, schematically, the areas of accretion and erosion in the form of curves of constant rate of bed-level change multiplied by 10^7.

References

1. Abbott, M., *Computational Hydraulics*, Pitman (1979)
2. Brebbia, C. A., *Computational Hydrodynamics*, Butterworths (1983)
3. Brebbia, C. A. and Walker, S., *Dynamic Analysis of Offshore Structures*, Butterworth (1979)
4. Bowden, K., *Physical Oceanography of Coastal Waters*, Ellis Horwood (1983)
5. Csanady, G., *Circulation in the Coastal Ocean*, Reidel Publ. Co. (1982)
6. Delft Technical U. Coastal Eng. Group, *Coastal Engineering*, Vol. I, II (1980)
7. Graff, *Hydraulics of Sediment Transport*, McGraw-Hill (1971)
8. Horikawa, K., *Coastal Engineering*, U. of Tokyo Press (1978)
9. Koutitas, C., *Elements of Computational Hydraulics*, Pentech Press (1983)
10. Le Blond, P. and Mysak, L., *Waves in the Ocean*, Elsevier (1978)
11. McDowell, D. M. and O'Connor, B. A., *Hydraulic Behaviour of Estuaries*, McMillan (1977)
12. Meyer, R., *Waves on Beaches and Resulting Sediment Transport*, Academic Press (1972)
13. Mitchell, A. and Griffiths, D., *The Finite Differences Method in Partial Differential Equations*, Wiley Int. (1980)
14. Ramming, H. and Kowalik, Z., *Numerical Modelling of Marine Hydrodynamics*, Elsevier (1980)
15. Rodi, W., *Turbulence Models and their Applications in Hydraulics*, IAHR (1980)
16. Roache, P., *Computational Fluid Dynamics*, Hermosa (1976)
17. Silvester, R., *Coastal Engineering*, Elsevier (1974)
18. Smith, G., *Numerical Solution of Partial Differential Equations*, Oxford (1986)
19. Sorensen, R., *Basic Coastal Engineering*, Wiley Int. (1978)
20. Svendsen, I. and Jonsson, I., *Hydrodynamics of Coastal Regions*, Techn. U. of Denmark (1980)
21. Strang, C., *Introduction to Applied Mathematics*, Welseley Cambridge Press (1985)
22. U.S. Army Corps of Engineers, *CERC Shore Protection Manual*, Vol. I, II, III (1973)
23. Wood, M. and Flemming, J., *Coastal Hydraulics*, McMillan (1981)

Index